PARTNERSHIP, COLLABORATIV
URBAN REGENERA

T0227931

Urban and Regional Planning and Development Series

Series Editors: Professor Peter Roberts and Professor Graham Haughton

The Urban and Regional Planning and Development Series has developed a strong profile since it was launched in 1995. It is internationally recognised for its high quality research monographs. The emphasis is on presenting original research findings which are informed by theoretical sophistication and methodological rigour. It is avowedly global in its outlook, with contributions welcomed from around the world. The series is open to contributions from a wide variety of disciplines, including planning, geography, sociology, political science, public administration and economics.

Other titles in the series

Partnership, Collaborative Planning and Urban Regeneration

JOHN McCARTHY
Heriot-Watt University, UK

Routledge
Taylor & Francis Group

LONDON AND NEW YORK

First published 2007 by Ashgate Publishing

2 Park Square, Milton Park, Abingdon, Oxon OX14 4RN
711 Third Avenue, New York, NY 10017, USA

Routledge is an imprint of the Taylor & Francis Group, an informa business

First issued in paperback 2016

British Library Cataloguing in Publication Data
McCarthy, John, 1961-
    Partnership, collaborative planning and urban regeneration.
    - (urban and regional planning and development)
    1. Urban renewal - Scotland - Dundee 2. Urban renewal
    3. City planning - Scotland - Dundee 4. City planning
    5. Dundee (Scotland) - Economic policy
    I. Title
    307.1'216'094127

Library of Congress Cataloging-in-Publication Data
McCarthy, John, 1961-
    Partnership, collaborative planning and urban regeneration / by John McCarthy.
        p. cm. -- (Urban and regional planning and development series)
    Includes index.
    ISBN 978-0-7546-1375-6
    1. Urban renewal--England. 2. Urban renewal--Scotland--Dundee. 3. Urban renewal--Government policy. 4. Urban policy--England. 5. Urban policy--Scotland--Dundee. I. Title.

    HT178.G72E5458 2007
    307.3'4160942--dc22

                                                                        2006103137

ISBN 13: 978-0-7546-1375-6 (hbk)
ISBN 13: 978-1-138-25802-0 (pbk)

Contents

List of Figures and Illustrations

Chapter 1

Introduction

Introduction

This book addresses the nature and complexity of urban regeneration policy and practice by means of a specific case study of the experience of regeneration in Dundee in Scotland. This introductory chapter sets out the broad frame of reference of the book by introducing the main cross-cutting themes and issues. First, aspects of urban decline are put within a global context and important themes are highlighted. Second, the institutionalist approach, which informs the development of ideas in the book, is set out. Third, important issues of policy transfer and learning are introduced, given the prevalence of such learning and transfer for urban regeneration policy between different contexts, particularly between the UK and Europe, as well as between the UK and the USA. Finally, the structure of the book as a whole is set out by means of a summary of the content of each chapter.

Urban issues in a global context

The issues underpinning problems of urban decline in a global context may be seen to arise as a result of changes within the dynamics of Western capitalism which have taken place in the second half of the twentieth century and to the present day. These changes have included the decline of manufacturing, associated growth in unemployment and social polarisation, and corresponding growth in social exclusion, albeit with an increase in employment in the service sector and knowledge-based industries in particular. A number of separate but interrelated factors may be distinguished in this respect.

First, the restructuring of employment within cities has arisen from the introduction of increasingly flexible and deregulated labour markets, together with global competitiveness and technological innovation. The result has been symptoms of decline in many cities, as characterized in Chapter 2.

Second, processes of globalization have themselves led to a reordering of the significance of cities, and new patterns of spatial relationship between areas of exclusivity and areas of exclusion have been established. The key characteristic in this context is that of 'uneven development', for instance within what have become known as 'two-speed' cities, and this is considered in more detail in Chapter 2.

Third, issues of urban governance have arisen as a result of shifts away from traditional models of urban government and the introduction of concepts of governance, partnership and 'entrepreneurial government'. This has presented a range of opportunities for solving or addressing problems of urban decline, but it has also raised important issues of accountability and inclusion (Jewson and MacGregor 1997).

The institutionalist perspective

The institutionalist perspective assists with the understanding of social, economic and political activity, within which the concepts and practices of urban regeneration policy are embedded. Essentially, this perspective involves the assumption that individuals do not exist as autonomous units in terms of their choice of actors, and then make rational choices in a linear fashion to maximize their interests. Rather, this approach acknowledges that we act and make decisions through the links we have with other people and through the values and norms that we acquire as a result of such interaction. This interaction takes place in the particular context of a world structured by power relations that derive from economic, political and other organizations. But within this context, people can act independently (Healey 2006). Thus, 'individual agency, in the ongoing flow of relations and actions, re-enacts and reforms power structures through acknowledging them or deliberately seeking to change them' (Healey *et al* 1995, 14).

This process – the interaction of structure and agency – underpins this book. The activity of 'structuration' assumes that 'People are ... embedded in their social relations, both shaped by them and actively constituting them through the routines of daily life and through deliberate strategies of relation-building' (Healey *et al* 1995, 14). This implies the need to consider 'cultural communities', 'discourses', 'networks', 'institutional capacity' (see Chapter 2) as well as the broadest use of power. The latter is particularly important since the exercise of power maintains other structural driving forces, though its effect is contingent on the particularities of context.

The linked notion of 'institutional thickness' refers to a set of factors including a strong institutional presence (with a mix of different institutions), a high level of interaction amongst local networks of institutions, and shared cultural norms and values. Together, these factors encourage the development of trust and co-operation which seem to be critical to the stimulation of entrepreneurship and economic development (Amin and Thrift 1995). The notion of institutional thickness is important in understanding and informing the efforts of local governing regimes to develop their local economies in an endogenous manner. Moreover, as Amin and Thrift (1995) point out, the development of institutional thickness is independent of the size of urban area, with the implication that a city such as Dundee can exhibit institutional thickness in a similar way to Edinburgh or London, although other contextual factors are relevant in applying this concept.

Nevertheless, determining precisely how such institutional thickness can be enhanced most effectively in order to develop competitive advantage is much more problematic. As Amin and Thrift (1995, 107) suggest:

> One of the distinctive features of less developed or disadvantaged urban areas is that their historical integration into the global capitalist project has stripped them of any coherent or cohesive institutional infrastructure as well as a common agenda which

might serve to develop structures of dominance. More often than not, these cities are locked into narrowly bounded alliances between powerful local politicians, successful local businesses and bodies representing cultures of 'colonisation'. Thus, while they may well be capable of developing supply-side initiatives and going some way towards inter-institutional collaboration, the key *cultural* foundations for extended, locally-rooted growth (e.g. "studied" trust, community-wide shared norms, distinctively local traditions, etc.) are likely to be more difficult to build.

Indeed, Amin and Thrift (1995) suggest that dominant institutional cultures may constitute an obstacle to change that could represent a threat to local development based on sectionalism. That this does not seem to have occurred in the case of Dundee may be explained in part with reference to the 'corporatist' context for public policy and its formulation and implementation within Scotland (see Chapter 4). It also reflects the historical legacy of intervention, including the evolving actions of bodies such as the Scottish Development Agency (SDA). The SDA fostered partnership between local agents in Scottish cities in the 1980s, which in the case of Dundee were sustained and enhanced with the development of the Dundee Partnership.

However, many attempts to force such institutional thickness by establishing new institutions and structures have not succeeded in ensuring that such institutions or structures are accepted and used by the various agents involved. Essentially the 'product' has often been provided without awareness of the need to embed the 'process' of development of institutional capacity, and the cultural aspects of partnership development have not been recognized. However, the application of concepts such as those above should not be confused with analyses of the urban problem that rely on an interpretation informed by economic determinism. This might involve for instance the assumption of the need for urban entrepreneurialism as the main response – in all circumstances – to processes of 'globalisation' and resulting economic restructuring (Graham 1995) (see Chapter 2). Such an explanation of the urban problem reflects a specific set of ideas and values. Clearly, differences in the institutional context of cities give rise to the need for different approaches to urban regeneration. This point is developed in several parts of the book, since Dundee is in many ways atypical in terms of its social, political and economic circumstances.

Moreover, a development of the above argument suggests that the need for institutional thickness that is locally-based need not be a precondition for the development of competitive advantage. This is because the case for local action in this respect has often rested on the perceived need to accept non-interventionist policy at the national level. Consequently, if institutional capacity and willingness to intervene in disadvantaged areas is enhanced at the national level, then the compelling need for action at the local level to enhance institutional thickness might be reduced (McCarthy and Newlands 1999).

Anglo-American and European perspectives

Given the importance of the need for understanding of local contexts in the framing of urban regeneration strategies and initiatives, it is clear that there is also a need for understanding of implications of differences in national political, social and economic contexts, where lessons and policies are transferred from one national context to another, as in the case of urban regeneration policy. Indeed it may be argued that such policy transfer has applied more in the case of policy for urban regeneration than for other types of public policy because, following the introduction of an explicit urban policy in 1968, the UK has in many ways been influenced by the USA in the development of such policy.

However, this has been in spite of major differences in context. Le Gales (2000), for example, highlights the potential differences in the US and European contexts in terms of private sector interests and public governance. He shows that concepts such as urban regimes and growth coalitions originated in the USA. Hence such concepts are more attuned to the US context, particularly in terms of the structural dependence of American cities on private business and the paramount importance of the driving force of property development in many contexts. He also makes clear that there are additional differences between the US and European contexts, particularly the following:

- a relatively limited role for private firms in the governance of UK cities;
- a relatively high degree of dependence of US municipalities on locally-raised taxation;
- a relatively high level of public ownership of land in the UK;
- a relatively high level of centralisation of financial institutions and firms in the UK;
- a relatively complex and developed network of relationship between local authorities and the state in countries such as the UK and France (Le Gales 2000).

Clearly, such differences must be taken into account in learning lessons and attempting policy transfer between different national contexts. Moreover, of course, relative to many other countries within the European Union, the UK appears since the 1980s to have been neo-liberal in orientation in many aspects of policy, with many public sector programmes having been reoriented towards forms of business sector involvement and engagement, private funding and a reliance on public-private partnership. Again, this implies that care must be taken in transfer of policy involving the UK.

Structure of the book

Following from the introduction in this chapter of the main themes to be highlighted in the book, Chapter 2 sets out the main theoretical frameworks that underpin much of the book. Here, the nature of the 'urban problem' is set out, together with its main consequences. This is followed by a consideration of the global and local aspects of the urban problem, in terms of both its nature and the responses that it has prompted. This leads to a consideration of critical issues of governance and the development of partnerships and networks. The framework afforded by the approach of communicative action, building on the institutional approach set out in Chapter 1, is also considered, together with the prism afforded by Giddens's typology of policy phases associated with the 'Third Way' concept.

Chapter 3 deals with the policy response to the urban problem in more detail, in the context of England. It considers the emergence of policy for urban regeneration, and sets out what have come to be seen as the key phases of urban policy since the 1960s. However, policy has developed in a different way in different contexts within the UK, and Chapter 4 sets out the parallel evolution of urban regeneration policy in Scotland. Specifically, this Chapter illustrates the role of the Scottish Development Agency (SDA) in setting an early framework for regeneration practice, and it shows the evolution of initiatives such as the Priority Partnership Areas (PPAs) and the Social Inclusion Partnerships (SIPs), culminating in the Community Regeneration Fund and community planning. This Chapter then sets out the dominant similarities and differences between urban regeneration policy approaches in Scotland as compared to England.

Chapter 5 introduces the case of Dundee by considering its economic, social and environmental context. It examines the economic background to both the city and its region, and then puts this in a historical perspective. The emergence of social and economic problems in the twentieth century is highlighted, together with the regional planning response of the Tay Valley and Dobson Chapman Plans. Regional industrial policy in the latter part of the twentieth century is also considered in terms of its effect on Dundee, particularly in terms of inward investment. This leads to consideration of the emergence of the partnership approach, as shown by the creation of the Dundee Project and later Dundee Partnership. The main institutional actors are also introduced, including Dundee District Council (now Dundee City Council), Tayside Regional Council (as was), and Scottish Enterprise Tayside.

Chapter 6 considers the problems and potential of the 'property-led' approach to urban regeneration as illustrated by the case of Dundee, and selected projects illustrate significant sub-themes. Hence an 'industry-led' approach is illustrated by the case of the Blackness Industrial Improvement Area; a 'tourism-led' approach is illustrated by the case of the Central Waterfront; and a 'culture-led' approach is illustrated by the development of the Dundee Contemporary Arts (DCA) project.

Chapter 7 considers a larger-scale example of the 'property-led' approach to regeneration, as applied to housing, in terms of the Whitfield Partnership, an important and seminal policy initiative for the regeneration of a peripheral estate on the outskirts of Dundee. This initiative was also of wider significance as one of the four Scottish Housing Partnerships. This Chapter sets out the origin and development of the Whitfield Partnership, and assesses its achievements and shortcomings.

Chapter 8 provides a contrast to Chapters 6 and 7 by considering what may be regarded as more 'holistic' approaches to regeneration incorporating a clear social dimension in particular. These are considered primarily in terms of the Social Inclusion Partnerships (SIP) initiative, as well as its precursor, the Priority Partnership Areas initiative, and its current equivalent, the Community Regeneration , as well as the broader role of the Dundee Partnership.

Chapter 9 builds on previous chapters by summarising the main achievements and outstanding problems in terms of Dundee's experience of regeneration, and it also considers the broader contested and unresolved issues and tensions in the area of urban regeneration. The main critical issues are clearly common to many other contexts within the UK, as well as for all cities attempting to cope with long-term restructuring. For instance, conflicts remain between modes of allocation of urban funding based on need as opposed to those based on opportunity, as well as between economic priorities and the need for 'holistic' approaches to regeneration.

Chapter 2

Theoretical Frameworks

Introduction

This chapter identifies the nature of the urban problem, its most important causes, and the explanatory theoretical concepts and frameworks applied in connection with this problem. First the broad nature of the problem is set out. Second, social effects, with associated explanatory concepts, are considered. Third, the different global and local aspects of problems of urban decline, and responses to them, are articulated. Fourth, spatial effects of 'globalization' are considered. Fifth, critical issues of governance are set out, with concepts relating to partnership and networks, and the theoretical frameworks afforded by notions of communicative action and associated approaches to governance and strategy development are considered. Finally, Giddens' typology of phases associated with the 'Third Way' concept, which helps to explain the trajectory of regeneration policy, is examined. This chapter therefore sets the context to understanding the dynamics of planning, development, regeneration and governance in Dundee.

The urban problem and urban decline

The magnitude of change in many of the world's cities is unprecedented, and processes of urban restructuring are re-shaping cities in ways unforeseen in earlier decades. One effect of such processes is what has come to be known as urban decline. In this context, the 'urban problem' may be defined as the relative under-performance of many local urban economies and the resulting mix of economic, social, physical and environmental exclusion, which often appears to be self-sustaining in the absence of external intervention. The result is that, without intervention, many urban areas appear to experience a self-sustaining downward spiral of decline in many respects. Such decline involves a variety of symptoms at the local level. For instance, social indicators include increasing income differentials, crime and racial conflict; and economic indicators include de-industrialization, manufacturing decline, increasing unemployment and welfare dependency, and infrastructural decay (Pacione 1997). These processes often reinforce each other, with the result that the incidence of disadvantage or exclusion becomes increasingly concentrated.

The causes of urban decline are more contested and difficult to determine. Certainly, structural factors linked to broad social and economic change and the globalization of economic activity contribute significantly to the local incidence of decline. However, local factors, such as the operation of local governance in responding to the broader restructuring of the state and the market, may also be significant in this context. The result is frequently an increasing incidence of poverty and multiple deprivation or disadvantage. The underlying reasons for the emergence of what may be seen as the 'urban problem' are therefore complex and interconnected.

It is useful now to draw out in more detail the way in which economic restructuring has contributed to the urban problem, and the response in terms of developing theoretical frameworks.

Post-war restructuring of the UK economy

The so-called 'long post-war boom' in the UK meant that between the early 1950s and the early 1970s labour productivity and wages at least doubled, with the fastest growth rates experienced in the new industries such as vehicle manufacture and chemicals and petroleum products. The expansion of manufacturing during this period was accompanied by parallel growth in the range and scale of output of personal, business and public services. However, in spite of the overall buoyancy of the economy, with full employment and rising prosperity, older industries such as shipbuilding, coal mining, iron and steel production, and heavy engineering, were all undergoing serious problems of contraction. Furthermore, the economy as a whole began to experience problems of reduced international competitiveness as shown by declining exports, and by the early 1970s the indications were that the post-war boom was ending (Pacione 1997), with significant regional disparities becoming more evident. The focus was therefore on the regional dimensions of change in economic activity and employment.

An important feature of the end of the boom was the de-industrialization of the manufacturing base of Britain's cities. This at first consisted of relative decline as service sector growth outpaced that of manufacturing, but later it took the form of absolute decline as manufacturing employment was reduced – for instance by more than one million jobs between 1966 and 1976 (Pacione 1997). Most manufacturing sectors were affected; for instance the loss of manufacturing employment in the West Midlands led it to become known as Britain's 'rust belt'. While in the latter case an important factor was that of intensive overseas competition, it may be argued that this general process of de-industrialization was exacerbated by government policies in the 1980s. This was because the anti-interventionist, free market based philosophy which prevailed at this time appeared to worsen the local effects of wider processes of economic recession. Of course, the effects of this were consequently felt disproportionately on those areas largely dependent on manufacturing, namely peripheral regions such as Tyneside and Glasgow, but important effects were also passed on to those industries dependent on manufacturing, such as coal mining.

The loss of employment in many manufacturing cities was matched by a decentralization of population as people began to move out to suburban locations, and between 1951 and 1981 the largest cities lost on average one-third of their populations. Clearly, the effects of such major shifts were worsened by the selective nature of the population movements. Hence those most likely to move were the younger and more affluent who left by choice, while those who remained were generally groups such as the elderly, young adults and others who were usually living on below average incomes, and who did not have the opportunity to move away from the city. In some cities, newly vacated areas were occupied by immigrants who were subject to compounded disadvantage since social and economic problems were exacerbated by racial discrimination. This was the origin of the so-called 'inner city problem' (Pacione 1997).

Regulation theory

As a result of the kind of economic transformation indicated above, regulation theory emerged as a means of responding to the crisis of post-Second World War Keynesian economics during the 1970s. The regulationist view challenged the previously dominant Keynesian view, which had assumed that economic crises could be managed by fiscal policy and control of the aggregate level of demand. Instead, the regulationist view proposed that the problem lay with the inability of capitalism to ensure sustained accumulation. In this context, it is important to note the shift from a 'Fordist' to a 'post-Fordist' mode of production, with the move away from the era of mass industrial production in both the USA and the UK. Hence, by the 1970s, the increasing use of flexible specialization within industry, facilitated by technological advances and changes in industrial organization, meant that companies needed to more effectively cater for niche markets and become more responsive to market conditions. This implied the need for a move away from centralist approaches to government towards more innovative alternatives. However, this approach emphasized the positive outcomes of market behaviour, and it may be criticized for underestimating the negative effects in terms of neglect of social provision (Jacobs 1992).

It is now appropriate to examine the social effects of the urban problem in more detail, together with the detailed concepts that have emerged to explain the nature of these effects.

Social effects of the urban problem

As a result of the nature of the perceived urban problem described above, problems of poverty increased in many of Britain's cities. Indeed, during the 1980s, poverty was seen to increase faster in the UK than in any other EU member state, and the gap between rich and poor also widened. Again, the geography of this incidence of poverty was markedly concentrated, since a substantial proportion of the most disadvantaged lived in Britain's cities – particularly those which had been subject to significant manufacturing decline and which had not experienced revival since they were not seen as profitable locations for capital. The problems faced by those still living in such areas were complex and interconnected – hence the term 'multiple deprivation' was used to describe the condition whereby a range of social, economic and environmental aspects of disadvantage compounded each other (Pacione 1997). Such multiple deprivation appeared to be concentrated in particular areas of cities. This spatial concentration added significantly to the problems of those living in such areas, who suffered from a heightened incidence of unemployment, crime, delinquency, morbidity, poor health, underachievement, and drug abuse.

Unemployment may be seen as the dominant causal factor within such a list of exclusion. This is because the lack of access to employment tended to lead to a dependence on social welfare and lack of disposable income that was often accompanied by lack of self-esteem, poor health and clinical depression. Again, however, certain groups suffered more than others. In particular, ethnic minorities appeared to suffer more since they tended to be disproportionately concentrated in those wards with the highest levels of deprivation, particularly in terms of unemployment. The reason for this are linked to the fact that migrants had been actively recruited to meet labour shortages in manufacturing industries and public services. These jobs often had low wages and minimal employment rights, and the restructuring of the economy outlined above therefore led directly to widespread unemployment amongst ethnic minorities in particular. This was compounded by government policies to restrict employment rights and to introduce reductions in welfare support, in the 1980s in particular. Such problems may be reinforced by the spatial concentration of such groups, often compounded by housing allocation practices and broader discrimination (Pacione 1997).

The urban 'underclass'

A particular concept used to describe the way in which the incidence of poverty and deprivation has changed is that of the 'underclass'. As a result of processes of social and economic polarization and marginalization, there may be said to have emerged an 'urban underclass' consisting of 'those forced to the margins or out of the labour market by the advent of advanced capitalism' (Pacione 1997, 54). This 'underclass' therefore comprises people who do not share in the benefits of growth enjoyed by the rest of society. The members of this group suffer from persistent poverty, act in ways that are contrary to the rest of society, tend to be spatially concentrated, and may be seen to form part of a sub-culture which means that behaviour is transmitted to the next generation. Of course the presence of such groups, and the socio-spatial divisions which bring them about, may be regarded as an inevitable part of uneven capitalist urban development. However, there may be seen to be an advantage to the rest of society in the increase in social cohesion that results from ameliorating the conditions of such an underclass. This is because over a long period such a group is likely to manifest itself upon the rest of society in the form of increased crime and broader social unrest. In fact, the latter may be seen to be the trigger for the emergence of, and the development of specific forms of, urban regeneration policy, as indicated in Chapter 3.

It is important in this context to note the varying explanations of the emergence of an urban underclass. Two basic variants of such explanations may be identified. The first tends to stress the causal importance of the characteristics of the members of the underclass. Hence, behavioural traits of the members may be identified in terms of unemployment, criminality, and high incidence of unmarried mothers, for instance. This is clearly related to the 'culture of poverty' thesis promoted by Lewis (1968), since it assumes that those born within such a culture come to absorb its basic attitudes and adopt its behavioural norms and practices. They are therefore, it is argued, unable or disinclined to take advantage of changing conditions or to identify opportunities for advancement. This implies that they may not have a desire to seek out employment opportunities in the way others would seek to do. Moreover, several observers such as Murray (1984) have proposed the view that the welfare state may exacerbate the problems of such people since it may create a so-called 'dependency culture' whereby the unemployed come to depend on state benefits rather than on paid work. Murray's ideas originated in the USA, but he applied his ideas to the UK and suggested that the size of the 'underclass' in the UK could proportionately become greater than that in the USA.

The second type of explanation stresses the importance of structural factors, particularly the failure of the economy to generate sufficient jobs to absorb an expanding supply of labour. This, it may be argued, leads to long-term unemployment, with inadequate welfare provision often contributing to a commensurate increase in the incidence of poverty. Clearly, the first type of explanation implies that the origin of the problem lies with the individual or groups suffering from poverty or exclusion, while the second points to the broader economic and political structures as the basis of the problem. Interestingly, the emergence in the 1980s and 1990s of legislative attempts to persuade the unemployed to seek work and to coercively withdraw benefits where this was not done would seem to reflect a shift towards a reliance on the first type of explanation. This is perhaps ironic given that it is precisely this explanation that gave rise to the first wave of urban policy initiatives in the 1960s, as indicated in Chapter 3, but which was later discredited as a result of the evidence arising from these very initiatives.

Social exclusion

The emergence of the concept of social exclusion in the 1990s reflected changing perceptions of poverty and disadvantage (Cameron and Davoudi 1998). In particular, there was at this time a move away from perceiving poverty and disadvantage as a 'static concept, an outcome rather than a dynamic process ... mainly concerned with income distribution and defining what constitutes an adequate level of income' (Cameron and Davoudi 1998, 253). This was because of the development of a more critical appreciation of the complexities of social and institutional restructuring, disempowerment and community capacity building.

Nevertheless, it may be argued that concepts of social inclusion and exclusion are characterized by ambiguity. Certainly, a confusing diversity of terminology is now evident. Hence associated terms may comprise, on one hand, negative concepts of social isolation, marginalization, segregation and fracture, and on the other hand, positive concepts of integration and solidarity (Allen 1998). Furthermore, different types of exclusion may be identified, including those associated with economic, social and political processes. Moreover, in addition to problems of ambiguity, terms such as social exclusion may be considered to reflect value judgements rooted in a particular intellectual and cultural tradition, as well as the assumption of a particular welfare regime (Mandanipour 1998). However, in the late 1990s a broad consensus emerged on the relevance of these terms at the spatial level. This consensus reflected a recognition that lack of access to employment was often the dominant indicator of exclusion (McGregor and McConnachie 1995). Nevertheless, it may be argued that a more sophisticated understanding of the dynamic processes of exclusion and inclusion is required to enable the urban problem to be tacked effectively.

As indicated above, the loss of manufacturing industry arising from the end of the post-war economic boom was a major factor leading to the emergence of the problem of urban decline. However, it is important here to disaggregate the dimensions of the debate in terms of the interaction between global and local problems and effects. Indeed, it may be argued that the reasons for the loss of manufacturing industry referred to above stemmed not only from technological change but also from the relocation of industry to newly industrializing countries with cheaper labour and other costs of production. This was exacerbated by the global mobility of capital, which made it easier to transfer work to cheaper labour sources. Hence many of Britain's cities have largely ceased to be centres of production in the way they once were; instead, they have become centres of consumption, incorporating for instance tourism and culture-related activities as well as administration functions. While new consumption and administration-related uses have of course created new jobs, these have tended to favour certain groups rather than others, with those not so favoured including for instance unqualified school leavers (Lovering 1997).

Unfortunately, groups such as semi-skilled or unskilled males, which have felt the brunt of the increased unemployment in many older manufacturing centres, are clearly those for whom many of the new uses held least promise in terms of access to employment. Again, where such people are also members of an ethnic minority, these effects have often been compounded by discrimination. This of course links with the concept of 'underclass' discussed above since it is precisely such groups that are most prone to joining such an underclass, as the unemployed tend to lose what skills they had and to lose motivation to seek work. Such groups may include older males who may experience difficulties in adapting to new employment sources. Increasingly, however, younger people who have never experienced a job are also disadvantaged because they lack the necessary skills to make themselves employable. The decline in male employment in many old manufacturing cities has exacerbated problems of family structures which have led to an increased incidence of single motherhood. It might seem that, in line with the 'culture of poverty' thesis, this exacerbates the lack of employability of the next generation since it tends to increase the number of children growing up in poverty as well as to diminish the prospects of such children benefiting from an appropriate mix of role models. Moreover, the increase in dissatisfaction amongst young males in particular is also likely to lead to increased crime, further exacerbating the problems of concentrated disadvantage since those who are able to move from such areas are more likely to do so (Lovering 1997).

These issues reflect the competing explanatory frameworks used above to explain the 'underclass', since on the one hand there are clearly international economic changes – in terms of global competition – that are bringing about urban restructuring. On the other hand, such competition is compounded by effects at the local level in terms of the formation of dysfunctional 'sub-cultures of poverty' which are in turn exacerbated by the concentration of poverty in many urban areas. Broadly, it may be argued that the 'cultural' arguments – centering on such concepts as the 'dependency culture' – are now in the ascendance. This might seem to imply the need for economic flexibility and geographical mobility in order to enable adaptation to the perceived economic realities of the new global order.

However, it may equally be argued that these arguments are oversimplified and need to be questioned, for several reasons. First, as Lovering (1997) indicates, such an approach privileges a particular set of interests who gain disproportionately from its application. Hence groups such as landowners and property speculators seek to argue that the only viable economic solution for Britain's cities is to bow to what are seen as prevailing economic pressures, by for instance seeking to attract international firms to new office locations. Second, it may be argued that there is an overly economistic conception of the tension between 'global' and 'local', with economic forces being seen to originate at the global level, and the local level being characterized by response in terms of either adaptation or resistance. This implies that the global dimension involves irresistible change, with the inevitability of production of an 'underclass' (Lovering 1997). As indicated below, this view is contended by many.

Locality and 'New Localism'

These circumstances led to the so-called 'locality debate' in the 1970s and 1980s. Many of the contributors to this debate sought to oppose the argument that the people most adversely affected by economic restructuring – namely the 'underclass' – were themselves to 'blame' for their plight. Instead, it was argued that the constant restructuring of capital implied an inherent tendency towards uneven development. The debate therefore centered on the political nature and significance of economic decisions, with firms choosing to locate in one area rather than another as a result of a mix of perceptions. The effects of this were seen to be the reinforcement of perceptions and the relocation of work from high wage to low wage areas. Where the debate parted significantly from the arguably simplistic and deterministic view associated with globalizing processes was in the assumption of the role of the state. Essentially, many observers argued that the state – particularly the local state – could have a much greater role than merely reacting to globalizing tendencies. Hence it was argued that the local state could become a forum within which a range of social interests could become involved in the shaping of frameworks for local economic development. The result could arguably comprise a shift in power away from capital to labour in terms of the motivating force for restructuring (Lovering 1997).

The so-called 'locality debate' led in turn to what has been called the 'New Localism' in terms of the policy approach of some local authorities to addressing issues of economic, social and physical decline. As a result, from the early 1980s, many local authorities prepared local economic development strategies that were often at the centre of corporate strategies. Within such strategies, the emphasis was commonly on seeking to attract and promote business, though distributional issues also often featured. Part of the rationale for such approaches was the realization that, as part of the 'globalization' processes referred to above, there was an increasing degree of competition between regions and cities, with the result that the key motors of the world economy came to be seen as city-regions. Such city-regions possessed key advantages that were assumed to be enhanced by local development strategies, which could be used to exploit forces of globalization.

Much research in recent decades has focused on the way in which such successful economic activity is 'embedded' in local cultural and social institutions. The implication is that a culture of trust and co-operation can be of benefit to prospects for long term growth within a context of global competition. Hence local governance can act as a positive force for economic development via the development of beneficial networks and partnerships. The public sphere can therefore help to build the very cultures of trust and co-operation that are now seen to encourage long-term growth and innovation. This would seem to endorse the 'New Localist' approach as a means of addressing problems arising from globalization.

Spatial effects of globalization

Areas of separation

The processes of globalization discussed above have given rise to specific spatial effects. These spatial effects may be felt in (a) production processes, and (b) the residential patterns that result from these shifts in production location. In terms of (a), it is useful to distinguish between what may be called higher-order and lower-order functions. At the higher-order end, the control functions of management, finance, law and politics are affected by globalization, though the centres of many cities remain important for such functions. At the other extreme, sectors such as cleaning give rise to largely unskilled or semi-skilled employment. While such sectors are partially dependent on higher-order functions, they are found in many places other than central cities.

Another spatial effect arises from the production of areas of concentrated social exclusion, in which 'race or ethnicity is combined with class in a spatially concentrated area whose residents are excluded from the economic life of the surrounding society, which does not profit significantly from its existence' (Marcuse and van Kempen 2000a, 19). In many cases, the creation of such areas is compounded or reinforced by the existence of exclusionary enclaves (often with physical embodiment for instance by means of gated communities), suburbs and 'edge cities', which may resist encroachment by perceived undesirable groups. Partly as a result of processes of globalization, the partitioning of cities in this way appears to be increasing.

In addition, a further set of locations may be identified in which the processes of globalization have a particular impact. These may be termed 'soft' locations, namely those which would appear to be ripe for change and new development. Such areas comprise for instance waterfront areas; centrally located manufacturing areas; brownfield sites; central city office and residential locations; central city amusement locations and tourist sites; concentrations of social housing; locations on the fringe of central business districts; historic structures; and public spaces (Marcuse and van

Kempen 2000b). Many of the areas of Dundee used as examples elsewhere in this book fall in to one or more of the above categories.

Uneven development

More generally, processes of globalization may be seen to lead to patterns of uneven development. Massey (1994) contends that the unequal relation of production in capitalist countries lies at the heart of this issue, with clear positions of dominance and subordination. As indicated above, this implies that some areas tend to monopolize control functions – the higher-order service functions referred to above – while others are assigned subordinate roles. The result is a spatial division of labour. For instance, in some areas in the north of Britain, there is a dominance of so-called 'branch-plants', which are often responsible to corporate headquarters outside the UK.

Partly as a consequence, the so-called 'dual city' has emerged, particularly in the context of the USA. While such as concept has been evident in many historical contexts, it re-emerged in the later twentieth century as a result of post-industrial transformation, uneven development, and the increasing marginalization of particular social groups, as indicated above. Again, a widening gap is becoming evident – expressed in spatial terms – between those who are able to take advantage of new opportunities for employment and those who are forced to rely on casualized employment or who remain unemployed (Mooney and Danson 1997). Such segmentation involves complex processes of change within cities, as illustrated by the case of Dundee.

Issues of governance

Entrepreneurial cities

As indicated above, many observers have argued that the actions of localities – and nations – can indeed have an effect on processes of globalization. This brings to the fore the way in which processes of governance can be organized so as to bring about local effects. In this context, governance may be considered as control over a range of local activities that does not just involve the state. The influential work of Osborne and Gaebler (1992) is important here. They suggest that, given the increasing influence of the global marketplace and the knowledge economy in which customers are becoming accustomed to high quality and wide choice, many bureaucratic institutions are increasingly inadequate. The need is therefore, they suggest, for a 'reinvention' of government, by allowing an entrepreneurial spirit to develop in the public sector. This may be linked with the parallel reduction in the redistributive and public investment roles of welfare states throughout the USA and Europe, as well as the challenge in all these contexts from international financial markets, again closely linked to processes of globalization. The overall result has been the perceived need for a reduction in the role of direct state production and the parallel growth of the so-called 'enabling' state (Jewson and MacGregor 1997), with clear effects in recent years on aspects of governance at all levels.

These processes may be associated with the approach to governance in the form of 'partnership', discussed in more detail below. This may be seen to represent a distinct form of 'urban regime' which has emerged in a very similar form in many cities throughout the world. The growth of public-private partnerships may be seen as part of an approach to urban entrepreneurialism which reflects the broader transformation of urban governance and which has given rise to entrepreneurial approaches to regenerating and 're-imaging' the city for external investors. Indeed, partnership is inherently attractive to government because it allows other interests than the state to contribute to objectives for regeneration, it spreads the responsibility for success and failure, and it allows the leverage of private investment to bring about public aims.

In this context, Harvey (1997) has identified the growth of the so-called 'entrepreneurial city' as the dominant response to urban problems, and it may be suggested that a consensus emerged in the 1990s throughout the UK on the need for endogenous local economic development initiatives. However, a number of variations were developed in particular local circumstances. For instance, on the one hand there were property-led strategies often dominated by business interests, and on the other hand neo-statist or even community-based initiatives. What all these had in common, however, was their fundamental dependence to some extent on market forces, and their concern to address the issue of urban or regional competitiveness. They therefore sought to secure specific competitive advantages for the city or region. It is significant here to note the assumptions that appeared to underpin the growth of the entrepreneurial city. This may be seen to be based largely on a 'geo-economic meta-narrative' which stresses the need to re-design economic strategies and institutions so as to prioritize wealth creation in the face of increasing global competition, since this is seen as the prior condition for social redistribution and welfare (Jessop 1997). These assumptions have been contested, as indicated previously. Moreover, in Scotland, the historical legacy of corporatism meant that a somewhat different approach was followed, as indicated in Chapter 4.

City competition

One result of the combination of the process of uneven development and the rise of entrepreneurial cities has been the development of forms of competition between cities and regions for investment, increasingly on a global scale. Consequently, many cities have sought to promote an image that it is hoped will attract investors on an international basis. Strategies may range from stressing the environment, educational facilities, cultural facilities, scientific capacity and prowess, heritage or any number of other features which are thought to be attractive or significant for potential investors. This has resulted in local authorities and other agents, often subsidized by central governments, promoting particular types of developments that are seen to contribute directly to such aims of 're-imaging'. For instance, tourist attractions, sporting facilities and conference centres are often seen to fulfill such a role (Jewson and MacGregor 1997). This is of course linked to with city marketing strategies which often make use of high-profile physical development schemes as a central feature.

Partnership and networks

The contemporary importance of partnership should be seen within the broad political context of the crumbling of the post-war consensus regarding the balance between the public and private sectors, and the resulting polarization of political ideology (Bailey *et al* 1995; Oatley 1988). In this context the perception of the 'urban problem' has changed as a result of the move from a liberal reformist approach to urban policy based on social provision and subsidy, to a more commercially oriented focus on partnership and leverage. More specifically, the development of partnership activity within the UK can be linked to the emergence of a post-1979 realization on the part of local authorities that 'a UK government, when it chooses, was constitutionally unassailable' (Harding 1998, 75). As a consequence, many local authorities previously antipathetic to market- and partnership-based activities adopted an approach of so-called 'new realism', within which concepts such as partnership with the private sector were embraced as the only realistic way of addressing how to meet their regeneration objectives within a climate of dwindling resources and capacity to effect change. As indicated in Chapter 3, this was strengthened by the adoption by government of the principle of partnership and the creation of partnerships as a qualifying criterion for consideration for competitive regeneration funding. Such a criterion was applied in the case of the English City Challenge and Single Regeneration Budget initiatives as well as the Social Inclusion Partnership (SIP) initiative in Scotland, and the need for demonstration of broadly-based partnership, including local communities, continues to be important within regeneration funding.

The notion of partnership has therefore become a central principle underpinning all aspects of urban regeneration policy in the UK, and Bailey *et al* (1995) suggest the presence of 'a broad consensus among the main political parties and practitioners that claims that partnership is now the *only* basis on which successful urban regeneration can be achieved' (1995, 1). However, the notion of partnership remains a relatively amorphous concept that can cover a variety of relationships, and partnership arrangements in practice vary widely in form, structure, and composition. This ambiguity inherent in the concept of partnership might even be convenient for many of its advocates, since it allows the concept to be adapted to a range of political and ideological circumstances. Hence, as Hall and Hubbard (1998, 6) suggest, such approaches 'offer something for all local governments, irrespective of political ideology. To the left, the entrepreneurial approach promises a way of asserting local co-operation, promoting the identity of place and strengthening municipal pride; to the right, it supports ideas of neo-liberalism, promotion of enterprise and belief in the virtues of the private sector'.

It is therefore necessary to determine more precisely what is meant by the notion of partnership. Bailey *et al* (1995) define partnership as 'the mobilization of a coalition of interests drawn from more than one sector in order to prepare and oversee an agreed strategy for the regeneration of a defined area' (27). It may be argued that 'partnership' can be used interchangeably with concepts of 'coalition' and 'urban regime', but this appears to obscure significant differences in the theoretical roots of these concepts. Moreover, a variety of relevant arrangements may be identified, including policy communities, territorial communities, professional networks, intergovernmental networks and producer networks. All these have the characteristic of a clustering of interests with a common focus. In this context, a coalition may be considered simply as a group of like-minded interests, possibly comprising those from the private and public sectors. Such a coalition may operate as a partnership, in that it may involve co-operation amongst the actors involved. Alternatively, if the activities of a coalition do not involve such co-operation, it may not be considered as a valid partnership. As indicated below, however, 'growth coalitions' must be considered separately, together with 'urban regimes'.

First, however, a useful distinction between 'partnerships' and 'networks' is made by Lowndes, Nanton, McCabe and Skelcher (1997), who suggest that, while 'partnerships' necessarily involve formal organizational relationships, 'networks' tend to be voluntaristic. This does not imply that networks are of less value in terms of facilitating regeneration outcomes. Indeed, it may be argued that it is precisely the 'fuzzy' nature of networks that allows them to be dynamic in operation. In addition, networks may provide the basis from which sustainable partnerships can arise. An important distinction between partnership and networks, however, arises from the fact that, while partnerships can be 'forced', for instance by the adoption of appropriate structures, networks rely on notions of trust, co-operation and mutual advantage (Lowndes *et al* 1997). Indeed, Lowndes *et al* (1997) suggest that the forced adoption of formal and forced partnership structures may actually reduce the potential for effective networking, but that effective networks can be facilitated by the joint working of individuals from different agencies within the same geographical area over a substantial period of time.

It is clearly relevant in this context to consider the essential motivation for partnership, and Mackintosh (1992) presents a useful framework for understanding the processes by which partnerships are constructed. Specifically, she proposes three main conceptual models for partnership that build on different aspects of motives for partnership. First, partners may seek to maximize 'synergy' by combining assets such as knowledge and resources, in order to achieve more together than they could individually. Second, partners may seek to bring about 'budget enlargement' by working together, so as to gain access to resources that would otherwise be denied. Third, partners may seek to achieve 'transformation' whereby some partners are exposed to different modes of working, resulting in the objectives and cultures of different partners converging towards a mutually acceptable, and potentially innovative, strategy. It may be argued, however, that, in the context of criteria regarding partnership structures being applied to finance from government, the motive of 'budget enlargement' is often dominant, with the potential loss of advantages arising from 'transformation'. In addition, to the extent that 'synergy' is also a factor, it may be suggested that what Hastings (1996) calls 'resource synergy', in terms of combination of resources, is more important than 'policy synergy', in terms of development of policy. However, it is 'policy synergy' that seems more likely to lead to the creation of potentially innovative and sustainable solutions.

Harding (1998) also examines the motives for partnership development, suggesting that organizations and interests enter into partnership for one of three reasons, with a resulting three-fold classification of partnerships. First, 'defensive partnerships' arise where an agency needs to secure the co-operation of other partners in order to continue its operation. Second, 'offensive partnerships' develop when an agency wishes to achieve things beyond its realm of competence. Third, 'shotgun partnerships' develop when a higher authority demands that an agency enters into a partnership, for instance where a funding body imposes the creation of a partnership as a funding criterion. The latter would seem to correspond to Mackintosh's concept of 'budget enlargement' in terms of partnership motivation. Furthermore, Bailey *et al* (1995) suggest that partnerships may unlock new land and development opportunities, bring about valuable place marketing and promotion, enable greater co-ordination of infrastructure and development, and facilitate local confidence-building and risk minimization.

Two further significant partnership concepts may be identified. First, growth coalitions stress the primacy of growth objectives among both private and public sector players; and second, urban regimes emphasize the greater involvement of private business interests with largely market-driven objectives, as a result of the fragmentation of governance. These concepts are considered in more detail below.

Growth coalitions

Growth coalitions represent one specific type of coalition that involves elite hegemony by means of the dominance of land-based interests that seek to encourage higher land values by means of urban growth; this is in effect a form of corporate-controlled regime. Molotch (1976) set out the original concept of growth coalitions, as applying to urban areas controlled by a land-based elite that competes with other areas. The 'growth coalition' concept assumes that growth is at the expense of other localities, and that a range of social, economic and political forces is incorporated in the growth coalition notion. As Molotch (1976, 310) states, 'desire for growth provides the key operative motivation towards consensus for members of politically mobilized local elites'.

The growth coalition idea focuses on how land and property markets determine how a locality is shaped. The strength of growth coalition theory therefore derives from its stress on the immobile factors that determine the essentially local nature of localities. The role of land and property, in terms of ownership and control, management and development, planning and regulation, is fundamental to the concept. The indicators of growth in this context include a constantly rising urban population, an expanding labour force, a rising scale of retail and commerce, more intensive land development, and increased levels of financial activity. While the concept assumes the presence of competition between land interests, it also shows how anticipation of potential coalition may constrain the intensity of local conflict. The concept also indicates how the operation of growth coalitions may bring about negative effects, including benefits that accrue to only a small proportion of local residents, and costs that are imposed at the local level. One result may be the emergence of a counter coalition (Lloyd and Newlands 1988).

However, the growth coalition concept has been criticized because of the difficulties involved in its application (Valler 1995), and because it fails to give adequate emphasis to interests other than those based on land and property (Wood 1976). Moreover, its applicability to contexts other than USA has been questioned. This is because of the presence in the USA of locally based systems of taxation and banking, as well as more locally based business arrangements. This means that there is more chance of local coalitions of land based interests developing, and that such coalitions are sustained by much more rigorous competition for investment between localities, than in contexts within Europe. In addition, the relative strength of local government in the UK and Europe means that in many areas the local state takes primary responsibility for local economic development, rather than coalitions of private capital (Hall and Hubbard 1998). However, within the UK, it may be argued that the 'growth coalition' concept may be more applicable to the corporatist characteristics of Scottish institutional relations (Imrie, Thomas and Marshall 1995).

Urban regimes

The concept of urban regimes was also initially developed in the USA, and it seeks to explain how changing modes of local governance impact on the operation of local partnership. It therefore considers the effects of the shift in the style of urban governance from a 'managerial' approach in the 1960s to a more 'entrepreneurial' approach in the 1980s (Harvey 1989). Specifically, regime theory suggests that 'Governance through informal arrangements is about how some forms of co-ordination of effort prevail over others. It is about mobilizing efforts to cope and adapt; it is not about absolute control' (Stone 1989, 5). Consequently, it illustrates how local government can act in concert with a wide variety of other actors in order to realize broad objectives. 'Regimes' refer simply to the informal arrangements by which public bodies and private interests function together in order to be able to make and carry out governing decisions (Stone 1989, 197). More specifically, an urban regime may be defined as a coalition of urban interests, including elected local government officials, that seeks to co-ordinate resources and generate governing capacity.

Urban regimes therefore address the presence of a diversity of interests and political agendas in the city. They also emphasize the political problems of building local partnerships or coalitions of private and public sector interests, the multiple objectives of these various interests, and the inherent and pervasive conflict between them (Cox and Mair 1988, 1991). However, the nature, composition, ideology, objectives and actions of a local regime will vary with the economic and political conditions of the local area concerned. In contrast to growth coalitions, which stress the dominance of land-based interests in creating a shared agenda for growth, urban regimes therefore assume a broader and more varied set of interests, and they assume local government to have a degree of autonomy in terms of action and ideology. Urban regimes would therefore seem to have more potential than growth coalitions to explain the operation of the broad yet stable local spatial alliances in cities such as Rotterdam (McCarthy 1998c). Consequently, Bailey *et al* (1995, 25) suggest that

'The relatively simple formation of the urban regime perhaps best describes the ways in which the local State needs to be located within its constellation of intersectoral interests'.

Again, however, both the growth coalition and urban regime concepts were originally developed in the USA, where very different modes of governance from those in the UK and Europe prevail. For instance, local government power has been traditionally more limited in the USA than in the UK, so that regime theory has perhaps more applicability in the UK context. Moreover, the early history of the United States led to a culture in which there was a '"civic expectation" that private initiative would serve communities through the development of non-governmental institutions' (Jacobs 1992, 198). Partly as a consequence, there is a wide variety of partnership and coalitional activity in the USA, in a way that is very different from the UK and Europe, though the gap may be narrowing as a result of globalizing influences and processes.

Communicative action

The 'communicative turn' in planning theory is of relevance to debates on theoretical frameworks in relation to urban decline as well as to urban regeneration activity. Central to this framework is Habermas' notion of communicative rationality. This notion argues that 'knowledge for action, principles of action, and ways of acting are actively constituted by the members of an intercommunicating community, situated in the particularities of time and place' (Healey 1996, 243). The implication is that urban regeneration strategies are formed by ways of acting that we can actively choose, after debate. Such ways of acting can help to develop stable patterns of consensus that in turn can inform and shape effective partnership arrangements for urban regeneration.

This approach has developed in part from institutionalist approaches to economic analysis, and the tools of institutional analysis – such as policy discourses, communities and arenas – can be used to focus critical attention on these issues. But it is concepts of 'agency' and 'structure' that are perhaps of most basic significance to the institutionalist approach. In this context, 'agency' helps to explain local governance relationships and the behaviour of stakeholders, and 'structure' – in terms of the patterning of social relations produced by organizations of economic production – helps to explain how contextual forces constrain decision-making processes. Regulation theory, which emphasizes 'structure', and urban regimes, which emphasize 'agency', are also of relevance here (Vigar *et al* 2000), as indicated previously. The institutionalist approach would seem to acknowledge complexity in terms of power relations, particularly in terms of the 'webs of relations' which allow interchange of information and development of consensus. It also suggests that such relational webs can be exclusive where they allow only privileged access to resources (Healey 1997).

It is on the basis of such concepts that the notion of collaborative planning is founded. This approach to strategy building focuses attention on the use of networks or the 'webs of relations' referred to above. These networks are embedded in past experiences, but the structuring power relations on which they are based are continually renegotiated, with the implication that such networks can be transformed. Essentially, such transformation can 'add value' so that 'institutional capacity', discussed below, is enhanced (Healey 1997). A particularly important concept in this context is that of discourse communities, which exist in specialist fields of knowledge production. It is the interaction of such communities that may be seen to formulate effective strategies for spatial planning or regeneration. The interaction between such communities involves discussion and searching for translative possibilities, as in the case for instance of public participation exercises concerning local plans. This represents a process of mutual learning resulting from a mutual search for understanding (Healey 1996). Healey points out that the practical application of such notions can be seen for instance in terms of the emphasis in management theory on group culture formation and empowerment, rather than management through hierarchical authoritarian structures. Again, this highlights the relevance of the communicative action approach to partnership formation for urban regeneration.

The issue of governance is clearly of relevance here. In this context, processes of governance, or 'the management of the common affairs of political communities' (Healey 1997, 59), themselves generate relational networks which can be sustained or transformed by governance processes. The concept of cultural communities is also useful here. This refers to communities with 'distinctive systems of meaning and ways of valuing and acting' (Healey 1997, 60). The institutionalist approach on which such concepts are founded gives particular emphasis to this notion of culture, specifically defined as 'the continuously re-shaped product of the social processes through which systems of meaning and modes of thought are generated' (Healey 1997, 64). In this sense, cultures provide 'ways of thinking embedded in a way of acting, while the way of acting is infused by a way of thinking'. The implication of such concepts in the context of institutional capacity building, discussed below, is that such activity creates a cultural community of its own which enables issues to be discussed and resolved more effectively (Healey 1997, 64).

In the context of urban regeneration activity, the additional value that can be derived from an institutionalist approach builds on the notion of the social resources available in different places. Hence different places can be seen to have different 'institutional capacities'. The insititutionalist approach can identify how 'added value' can be created by enhancing intellectual and social capital and thereby increasing institutional capacity by supporting and extending local networks and building further networks of trust and support. This may be referred to as 'institutional capacity building'. The notion of collaborative governance is of relevance here, and may be linked to notions of corporatism discussed above, which may be argued to be particularly relevant to the Scottish context.

It became clear in the 1980s in particular that business groups in many urban regions of the UK were seeking to build regional alliances with government agencies. While this comprises precisely the sort of collaborative governance and consensus building that is potentially productive, it is necessary also to take account of the need to spread power amongst the broader range of stakeholders involved, and to be inclusive in terms of all potential stakeholders. This highlights the potentially-exclusive effect of corporatist approaches which may be linked to notions of 'entrepreneurial consensus' that represent consensus building amongst key regional and local players. More equitable arrangements with greater potential for the sharing of power may be seen to derive from 'inclusionary argumentation' which seeks to blend the advantages of entrepreneurial alliances with the need to avoid tendencies towards elitist corporatism. The difficulties of achieving such a mix must not be underestimated, however, since these two tendencies represent a broader arena of continual struggle between 'techno-corporate' and 'pluralist democratic' approaches (Healey 1997).

It is also significant that the UK government's emphasis on strategy, policy integration and stakeholder involvement since 1997 has been combined to an extent with a localising and regionalising thrust in policy development, though it may be argued that this needs to be developed further. This suggests an opportunity to address issues of inclusionary argumentation and consensus building as outlined above. Specific ways forward in this context therefore include the development of new institutional capacities, the enhancement of social capital in terms of place-based networks and alliances, and the building of new forms of political capital such as city mayors, in conjunction with increased local autonomy. Such innovations can help to ensure that new forms of urban regeneration strategy building are both effective and inclusive (Vigar *et al* 2000).

'Third Way' policy analysis

In terms of the broad shifts in post-war approaches to public policy, including that for urban regeneration, Giddens (1998) provides some further illumination of underlying causes. Essentially, he provides a typology to illustrate the periods during which key political and economic ideas have prevailed since the Second World War, and this is of help in understanding the development of regeneration policy. The central assertion is that it is only by such a longer-term historical analysis that the current trajectory of policy can be understood. While Giddens' approach allies most directly to national economic policy, the latter does of course provide the key context for the ethos, funding and policy application of urban regeneration. Specifically, Giddens identifies the following phases as providing a basis for explaining the trajectory of public policy development.

Social democratic period (1945-1979)

The prevailing policy ethos or paradigm during this period was Keynesian in character, and it was associated with government intervention aimed at addressing market failure in labour and capital markets with the associated need to rebuild the national economy after the Second World War. This approach was reflected not only in the UK but also in many other European states that were facing similar issues and problems at the national level. The overall ethos was sufficiently broad to encompass both Labour and Conservative administrations in the UK, which all accepted the essential ideology of government intervention, albeit to varying degrees. There is an analogy here with what has been called the UK's 'post-war consensus', which prevailed until the late 1970s; this period is associated with the creation of the welfare state, intervention by government at national level in industrial sectors and labour markets, and national public expenditure programmes to create and sustain full employment.

However, during this period there was criticism of the social democratic approach's lack of impact on major economic issues such as unemployment, inflation and industrial restructuring. Hence there was gradual but cumulative interest in the alternative approaches presented by monetarism and 'supply-side' economics, with the associated assumption of the need to remove the obstacles to what might be seen as the efficient operation of market activity and resource allocation. These arguments in the UK became particularly prominent in the mid to late 1970s, culminating in the shift in government in 1979.

Neo-liberal period (1979-1997)

This phase is associated with the Conservative administrations in the UK led by Prime Ministers Margaret Thatcher and John Major. The ethos at this time was based on the promotion of market processes and outcomes in determining the allocation of resources and the resolution of conflict. This period is associated with 'Thatcherism' and the so-called 'New Right', and its effects were felt in all aspects of government policy, albeit to different degrees. Hence policies were modified to enable market forces to hold greater sway, and so de-regulation and privatization were promoted, and allied with a more general suspicion of any kind of government intervention. Paradoxically, however, this phase was also illustrated by increasing centralization in government, and in many respects the attempts to dismantle structures of intervention (including for instance the planning system) were largely unsuccessful.

Again, however, there were countervailing forces during this period, and the period of Prime Minster John Major illustrated a softening of the policy approach, as well as the beginning of a turn to greater inclusivity in regeneration approaches, as illustrated by the seminal City Challenge initiative as set out in Chapter 3.

'Third Way' period (1997-date)

This phase is associated with the approaches of the post-1997 New Labour governments, and it involves a pragmatic mix of elements from the social democratic and neo-liberal approaches. Essentially, this approach attempts to avoid the problems of, on the one hand, the bureaucratic, top-down government of the social democratic period (and in particular with the old political Left), and, on the other hand, the dominance of de-regulation and market processes associated with the new Right. In part, the aim was to show how social democratic principles could evolve to meet new challenges such as globalization, environmental concerns and the need for active citizenship, with recognition of the successes of the neo-liberal approach in some respects in relation to economic efficiency, productivity and competitiveness. Hence labour market flexibility for instance continued to be a crucial policy concern, though there was also a resurgence of urban regeneration activity and spending during this phase, which is associated with the plethora of new structures, institutions and spending programmes as set out in Chapters 3 and 4.

However, this phase has also been subject to criticism. For instance it is asserted by many that it lacks rigour and substance, being in essence a re-presentation of a fundamentally neo-liberal approach. This, many argue, is indicated for instance by the continued acceptance of the dominant need to accept globalizing processes and the associated need for competitiveness and labour market flexibility, albeit at the expense of elements of equity. Thus for instance the 'modernization' agenda in relation to spatial planning, closely linked of course to regeneration activity, is seen by many as representing the dominance of growth-led agenda, taking precedence over parallel aims for social inclusion and environmental amenity. Nevertheless, this approach, in terms of its application to regeneration, clearly accepts the need for inclusive approaches as well as the need for appropriate institutional capacity.

Conclusions

It is clear that the problems of urban decline are multi-faceted, and have emerged from a variety of interconnected factors, including global factors such as changing locations for production, and local factors such as institutional capacity and partnership. There are many explanatory frameworks that have been used to explain why urban decline has occurred, and many such frameworks would appear to be competing for legitimation, particularly those that emphasize 'globalization' and those that emphasize 'localism'. Nevertheless, such theoretical frameworks offer the key to the development of more effective regeneration mechanisms. In particular, frameworks relating to governance, use of partnerships and networks, and use of techniques of collaborative planning indicate the possibility of more inclusionary and effective approaches to strategy-building for urban regeneration. In addition, the typology of Giddens, highlighting the notion of the (albeit contested) 'Third Way' offers a means of understanding the evolution of urban regeneration policy in the UK since the 1960s.

Chapter 3

Urban Regeneration Policy in England

Introduction

'Urban regeneration policy' may be defined as policy aimed at the physical, environmental and social regeneration of areas suffering from concentrations of deprivation. This Chapter outlines the emergence and evolution of such policy in England, sets out the major phases of policy development, and identifies major points of transition. Finally, a series of issues and tensions within urban regeneration policy, which have persisted throughout all policy phases, is set out.

Emergence of urban regeneration policy

The Urban Programme

The General Improvement Areas established under the 1969 Housing Act were amongst the first initiatives aimed at addressing physical decline in cities in the UK (Carley 2000), but it is also important to highlight the role of a series of government-sponsored studies of inner urban area undertaken in the 1960s, highlighting the extent of social need in such areas, particularly in terms of housing and education (Blackman 1995; Lawless 1989). Partly as a consequence of these studies, mainstream inner city policy was brought into being in 1968 by the introduction of the Urban Programme (UP) and the Community Development Projects (CDPs).

The Urban Programme provided grant aid to local authorities faced with extreme instances of social need, urban deprivation and racial tension (Blackman 1995; Lawless 1989; Robson 1988). Initially, it was funded and administered by the Home Office, partly because the 'inner city problem' was seen as one of law and order as manifested by urban unrest in areas such as Notting Hill in London. Consequently, grant aid was initially geared to social objectives. Indeed, it may be suggested that the trigger for these early urban regeneration initiatives was the so-called 'rivers of blood' speech on the issue of immigration by Enoch Powell, a former Conservative minister, in 1968. While planning had been underway for some time on the Urban Programme and Community Development Projects, this speech appeared to act as a catalyst that prompted the Prime Minister, Harold Wilson, to take action. It is clear that the resulting initiatives had the advantage of benefiting an important part of the Labour government's core constituency at a time of crisis. However, the antecedents to the initiatives can be seen to come from the 'War on Poverty' programme in the USA.

The Urban Programme allowed local authorities to access part funding from central government for projects that could demonstrate 'special social need'. It applied positive discrimination in favour of selected groups and areas, and took the form of small-scale projects. The emphasis was on experimentation, self-help, co-ordination of existing services and rapid results, and there was little central control over the type of projects to be supported. While local authorities had to demonstrate 'special need', there was little definition of what this comprised. The resulting projects varied greatly, though they had in common their possible interpretation as a 'conflict displacement strategy designed to act as a pressure valve to prevent tensions in run down inner city areas from reaching boiling point' (Atkinson and Moon 1994, 6). The administration of the Urban Programme by the Home Office seems to support this view, particularly in terms of the concern for the need to manage problems arising from what was then seen as 'race relations'. The amount of funding initially allocated for the Urban Programme was limited to £22 million over four years.

Community Development Projects (CDPs)

Like the Urban Programme, these were essentially small area projects intended to focus limited resources on communities perceived as 'deviant'. For each local authority area identified, a specific area was selected and an action team appointed. In terms of funding the CDPs cost around £5 million over nine years. As in the case of the UP, local authorities contributed 25% of funds with the remainder coming from central government. The Home Office again acted as the co-ordinating body. Twelve CDPs were set up in mainly urban locations, though there were no clear criteria for selection, and the CDPs seemed to suffer from a lack of direction (Atkinson and Moon 1994). The CDPs effectively ended in 1978. While the CDPs were predicated on the assumption of a social pathology model of urban decline, with imputed solutions arising from self-help and mutual aid, many of the CDP teams subsequently rejected this model. Indeed, several went on to question the area-based approach itself, in favour of an explicitly Marxist explanation with uneven capitalist development at the root of the perceived problem.

Inner Area Studies and Comprehensive Community Programmes

Six Inner Area Studies (IASs) were set up in 1972. Their focus of concern was either on local government decision making and its impact on environmental problems (Oldham, Rotherham and Sunderland), or on the problems of small selected inner city areas (Toxteth in Liverpool, Small Heath in Birmingham, and Lambeth in London). The aim for all these was the understanding of the causes of urban decline. Like the CDPs, they concluded largely that the social pathology approach was inadequate, but instead of the Marxist approaches of the CDPs, they advocated a pragmatic approach that combined an area-based concern with an awareness of structural factors. The Comprehensive Community Programmes (CCPs), launched in 1974, represented an attempt to apply techniques of corporate management to the problem of urban decline. Two (in Gateshead and Motherwell) were formally launched, but by this time these were fully established. In 1978 the government had set out its White Paper as described below. Consequently, the CCPs were either sidelined or absorbed into this initiative, and the programme was wound up by 1980 (Bailey *et al* 1995; Atkinson and Moon 1994).

Evolution of urban regeneration policy

Underlying assumptions

Before considering the way urban policy has evolved since the 1960s it is appropriate to consider the underlying assumptions inherent within early initiatives. In fact all the urban initiatives originating during the period outlined above share several common characteristics in terms of basic assumptions. First, it was assumed that deprivation existed in defined areas, and so additional resources, co-ordination and targeting, it was assumed, could address the perceived dysfunctional nature of the areas concerned. Second, it was assumed that central government could implement initiatives based largely on the USA 'War on Poverty' programme in order to deliver remedial measures for these areas. Third, it was assumed that action research should be linked to such initiatives in order to enable full monitoring and possible wider application. Fourth, since these initiatives were essentially experimental, it was assumed that they should run only for a limited period.

Moreover, it is significant that the basic underlying conceptual model for all the early initiatives, at their inception, was the 'culture of poverty' thesis, with the implication that deviancy of spatially-defined groups and individuals lay at the core of urban decline. Furthermore, it is important to note that concepts of partnership did not form a major element of any of these early initiatives since results were understood to be obtained by improved planning, co-ordination and targeting by the public sector (Bailey *et al* 1995).

Policy after 1977

In a move away from the social origins of urban policy, an emphasis on economic and environmental objectives and programmes was introduced in the late 1970s (Blackman 1995). In particular, the 1977 White Paper 'Policy for the Inner Cities' (Department of the Environment 1977) widened the scope of the Urban Programme considerably, in terms of introducing a more significant role for the voluntary sector as well as greater targeting of resources (Cullingworth 2002; Deakin and Edwards 1993). In addition, the White Paper aimed to bring about a greater contribution from the private sector to urban regeneration, since it stated that 'local authorities ... will need to be more entrepreneurial in the attraction of industry and commerce' (Department of the Environment 1977, 33). As a consequence, a greater emphasis was placed after 1977 on industrial, environmental and recreational schemes, at the expense of those with a community or social focus (Atkinson and Moon 1994).

The 1977 White Paper also stressed the importance of partnership between the public agencies involved in urban regeneration, and it introduced an 'enhanced urban programme' with a hierarchy of eligible authorities comprising 'Partnership' authorities, 'Programme' authorities, and 'Other Designated Districts'. This hierarchy was intended to enable more effective targeting of resources to the areas of greatest need (Robson 1988; Lawless 1989). Hence most resources were focused on the 'Partnership' areas, which comprised the major conurbations containing such areas. However, problems of bureaucratic inertia and lack of integration, and the resulting inability to effectively co-ordinate the agencies involved, meant that the impact of the new funding regime on the serious problems of such areas was relatively marginal (Lawless 1989; Robson 1988). Furthermore, for the 'Other Designated Districts', the 'traditional urban programme' represented a smaller share of overall funding than that to which they had previously had access (Deakin and Edwards 1993). Significantly, the government began at this time to see the need for broadly-based partnership. While the White Paper asserted that local authorities were the most appropriate agencies to tackle inner urban problems, it also indicated that more collaboration was required between the government and the private sector as well as with local communities (Bailey *et al* 1995).

Policy in the 1980s

The period after the election of Margaret Thatcher in 1979 may be seen to constitute a very different approach to all aspects of urban policy. In particular, it brought a shift in the direction of regeneration policy, introducing a greatly enhanced role for the private sector (Blackman 1995; Lawless 1987; Robson 1988), and a greater emphasis on economic projects within the Urban Programme (Robson 1988). In addition, an increased priority was given to the physical, as opposed to social, aspects of regeneration. This was perhaps manifested most clearly by the introduction of Urban Development Corporations (UDCs), which took over local authority powers in several inner urban areas in order to promote property development-led regeneration. However, the problem of lack of co-ordination of urban policy, which had been identified in the 1970s, persisted after 1979 (Lawless 1989). This was reflected in the 1988 'Action for Cities' statement (HMSO 1988), which declared the government's intention to integrate existing urban policy mechanisms, albeit in the absence of extra resources (Deakin and Edwards 1993).

After 1988, the hierarchy of local authorities which could bid for grant aid was phased out in favour of a single Urban Programme designation, for which 57 local authorities were eligible (Lawless 1989). A series of other grant aid initiatives was also introduced in the 1980s, though many of these were subsequently amended. For instance, the Urban Development Grant (UDG), introduced in 1982, was seen as having a 'pepper-pot' effect, which spread resources too thinly. Consequently, it was replaced in 1987 by the Urban Regeneration Grant (URG). The 1988 'Action For Cities' statement subsequently simplified arrangements for grant aid by the introduction of City Grant, which replaced the UDG, URG and the Derelict Land Grant in urban areas (Lawless 1989). City Grant was intended as a more direct means of stimulating private sector urban investment, since it was paid by central government direct to the private sector. Hence local authorities, which had previously received, and contributed to, grant aid, were by-passed to an extent by City Grant, which therefore represented a continuation of the 'progressive exclusion of local government from urban regeneration initiatives during the 1980s' (Atkinson and Moon 1994, 208). However, priority was given to applications for City Grant within the 57 local authorities eligible for UP funding (Blackman 1995; Lawless 1989).

Other initiatives introduced in the mid-1980s included the Estates Action Programme (1985), which involved councils bidding to the (then) Department of the Environment (DoE) for funding for the improvement of council housing estates; City Action Teams (1985), which were intended specifically to improve the co-ordination of civil service departments in relation to particular inner city areas; and Task Forces (1986), which were geared particularly to the development of enterprise in small areas within the inner city (Lawless 1989; Blackman 1995).

Problem of co-ordination

In spite of the kind of initiatives outlined above, the problem of co-ordination in relation to urban regeneration policy persisted in the 1980s. This was highlighted in 1989 by the Audit Commission, which commented on the perceived 'patchwork quilt' of urban regeneration initiatives (Audit Commission 1989). The fragmented nature of the package of initiatives was due in part to the highly departmental nature of central government in England, a problem which the City Action Teams had been set up to solve in particular local areas. This problem of departmentalism – what has been called the 'silo mentality' – was exacerbated by the fact that no single government department had overall control of inner city policies. Moreover, the inner city responsibilities of several departments had shifted; for instance, the administration of the Urban Programme had been passed from the Home Office to the Department of the Environment, and the Departments of Trade and Industry (DTI) and Employment (DoEmp) also had a significant input into urban regeneration policy. This led to conflict between departmental objectives. For instance, while the Department of the Environment was attempting to focus investment in inner city areas, including several in inner London, the Department of Trade and Industry was encouraging investment away from London to peripheral regions (Keating and Boyle 1986; Lawless 1989). While the 'Action for Cities' initiative was intended to solve such problems of co-ordination, it seemed to be largely unsuccessful in this respect (Lawless 1989).

Policy in the 1990s

City Challenge: competition and co-ordination

Partly as a consequence of the problems of co-ordination experienced in the 1980s, the City Challenge (CC) initiative was introduced in 1991. This initiative addressed the issue of co-ordination by encouraging local authorities to establish partnerships directly with the private sector, voluntary organizations, Training and Enterprise Councils and other agents responsible for regeneration (Cullingworth 2002). In order to facilitate the involvement of the private sector in particular, the

City Challenge initiative encouraged local authorities to adopt fast-track decision-making procedures so as to ensure flexibility and responsiveness. It also encouraged a corporate approach to ensure that schemes would be part of a co-ordinated and strategic programme (Oatley 1995). Moreover, City Challenge encouraged the use of participative techniques involving local people, partly because of the experience of the early Urban Development Corporations, which had often provoked antagonism amongst local communities. In many ways, therefore, City Challenge represented a watershed in the application of urban regeneration policy in the UK. Perhaps the most important aspect of the initiative, however, was its emphasis on competition since it applied an extremely competitive regime for funding allocation based on bidding procedures. In the first year, eleven bids out of twenty-one were successful, and they were allocated an average of £7.5 million per annum over a five-year period. A further round was launched in 1992, with twenty winners out of fifty-four. Significantly, City Challenge funding was not additional, and the government expected that they would be supplemented by additional resources from other sources, including the private sector (Atkinson and Moon 1994).

Nevertheless, the City Challenge initiative was criticized by many observers. For instance, Oatley suggests that the City Challenge bidding process 'had no objective relationship to levels of need or even ability to deliver' (1995, 12). Moreover, resources for City Challenge were 'top-sliced' from existing different programmes such as the UP, City Grant, Derelict Land Grant, City Action Team and Estates Action (Blackman 1995; Cullingworth 2002). Hence, in spite of the improvements in local consultation and partnership within the City Challenge initiative, the 1994 DoE report *Assessing the Impact of Urban Policy* (Department of the Environment 1994) concluded that co-ordination within local authority departments and central government departments, as well as between these layers, was still seriously lacking. On a broader basis, the report concluded that the 'Action for Cities' initiative had not broken down compartmentalised grant regimes for urban regeneration. It criticized the capricious nature of urban policy, indicating that 'the evolution of policy for the cities during the past decade was regarded [by one of the consultees] less as an iterative process than a series of policy oscillations' (Department of the Environment 1994, xii). Partly as a consequence of such criticism, there were only two rounds of the City Challenge programme before its resources were incorporated into the new Single Regeneration Budget (SRB).

Single Regeneration Budget: consolidation

As indicated above, the City Challenge initiative did not solve the deeply entrenched problems within urban regeneration policy. However, the competitive bidding process introduced in the scheme was seen to be effective in encouraging an enterprise culture (Oatley 1995). Consequently, it was considered that this bidding process should be extended, and it seemed that a more coherent and integrated approach could be achieved by involving regional budgets and locally co-ordinated, area-based strategies. As a result, the government introduced the Single Regeneration Budget, which brought together 20 previously separate funding programmes in England, and which was based on competitive bidding to central government for funds. The

SRB was administered by the new Integrated Regional Offices of the Department of the Environment, and £1.4 billion was allocated for the SRB in 1994/5, with most being allocated for programmes based on land and property regeneration (Blackman 1995).

The aims of the SRB were similar to those of later phases of the Urban Programme, in that it aimed primarily to encourage employment by using public funds to leverage investment from the private sector. Crucially, however, several features of the City Challenge programme were also applied as part of the SRB. For instance, the downgrading of need as a criterion for grant aid, in favour of development potential and proven competence to achieve implementation, was evident in both programmes. However, the most important link with the City Challenge initiative was that of the process of competitive bidding, since the SRB involved individual bids being put to the Regional Offices, which in turn produced bids for regional funding to the Ministerial Committee for Regeneration. This Committee comprised 10 cabinet ministers and the Inner Cities Minister (Blackman 1995), and it co-ordinated overall needs and priorities, and advised the Secretary of State for the Environment, who was accountable to Parliament (Blackman 1995). The SRB therefore replaced the Urban Programme as the main vehicle for inner area regeneration in England, and the UP began to be phased out in 1992/3, with existing programmes being allowed to taper. As far as new projects were concerned, the UP was subsumed into the SRB from 1994/95.

The SRB represented a substantial progression from previous initiatives in terms of flexibility of response (Deas and Harrison 1994; Hill and Barlow 1995). However, it represented an overall reduction in the level of resources available to previous Urban Programme areas (Hall 1995), since it was potentially available to all local authorities rather than just the 57 Urban Programme authorities (Hill and Barlow 1995). Moreover, many smaller authorities appeared to be disadvantaged in the bidding process, because of the resources required to mount an effective bid (Hall 1995). Indeed, it was estimated that around 20% of bidding authorities in the first round spent more than £6,000 in preparing the bid. Furthermore, in many cases there did not seem to be sufficient time available in the bidding schedule to enable effective partnerships to be formed (Bennett 1995). Perhaps most significantly, however, the emphasis on the quality of bids themselves, rather than the underlying basis of need, posed a serious problem according to some observers (Hill and Barlow 1995).

A further problem was presented by the lack of accountability arising from the reduction of the role of local authorities in the bid-making process. While some three-quarters of the winning bids were local authority-led, the necessity for leveraged funding from the private sector to support the bid seems to have contributed to a marginalization of the role of local authorities (Deas and Harrison 1994; Hall 1995; Hill and Barlow 1995). The voluntary sector, which previously had a significant role via the Urban Programme bids submitted by local authorities, also saw its role reduced by the SRB, since only around three per cent of the total SRB fund was allocated to partnerships led by the voluntary sector (Nevin and Beazley 1995). As a result of such factors, the SRB seemed to bring about a shift away from aid to the most deprived areas towards those with demonstrable potential (Robson 1994). It was therefore suggested by many observers that SRB needed to be more sensitive to the needs of local communities, with Shiner and Nevin (1993) for instance indicating the need for a national agency for community regeneration, supported by local community regeneration units within local authorities.

English Partnerships: property

'English Partnerships' (EP), originally referred to as the Urban Regeneration Agency, was created to work in parallel with the SRB by promoting urban regeneration focused on development or redevelopment. EP was established by the Housing and Urban Development Act 1993, and it extended certain elements of the UDC model, since EP was intended essentially as a flexible, fast-moving, roving development corporation. In particular, it aimed at 'promoting job creation, inward investment and environmental improvement through the reclamation and development of vacant, derelict and underused or contaminated land and buildings' (English Partnerships 1994, 2). English Partnerships was granted extensive powers that included compulsory purchase, control of development and provision of road infrastructure. However, the government's aim was that EP would work in partnership with local authorities, with implementation carried out by the private sector (Blackman 1995). English Partnerships funding was not intended for projects on greenfield sites, unless they were linked to a specific initiative such as the attraction of inward investment, and funding was not normally available within UDC areas (English Partnerships 1994).

English Partnerships was therefore originally conceived as a kind of 'English Development Agency' along the lines of the Scottish Development Agency, and it was originally proposed that English Partnerships would take over responsibility for the UDCs and Enterprise Zones. However, the proposed role of EP was amended after the 1992 general election (Atkinson and Moon 1994). Consequently, its terms of reference changed to include the requirement that individual projects must be shown to fit in with a coherent and integrated regional strategy. Hence EP subsequently assessed bids in conjunction with the Integrated Regional Offices which administered the SRB. Significantly, however, EP brought no additional funding, since its Investment Fund replaced the previous allocations for Derelict Land Grant and City Grant (English Partnerships 1994). Moreover, while EP was accountable to Parliament via the Secretary of State for the Environment, this indicated a reduction in accountability since EP gained powers which were removed from local authorities (Blackman 1994). The limited funding available to the agency, combined with its concentration on facilitating property development, implied that its achievements were likely to be marginal (Atkinson and Moon 1994). Certainly, it seemed unable to provide a powerful co-ordinating role similar to that of the SDA in Scotland from 1976 to 1992.

Nevertheless, in 2006 English Partnerships was seen by the government as an important vehicle for ensuring the delivery of high quality sustainable growth, in a manner that is consistent with the broader objectives of the Regional Development Agencies, the Housing Corporation and the aim of the Department for Communities and Local Government (DCLG: formerly the Office of the Deputy Prime Minister [ODPM]) for inclusive and sustainable communities. The aim of EP is to concentrate on the earlier stages of major projects, with other partners making their contribution at a later stage. The priority areas for the activity of EP in 2006 were seen as areas of greatest need comprising the following: the four Growth Areas in the 'greater south east' (namely Milton Keynes, the London-Stanstead-Cambridge M11 Corridor, the Thames Gateway and Ashford); the coalfields; strategic brownfield sites in or adjacent to the latter areas or in areas of housing pressure or low demand; and the 20% most deprived wards in England. The future activity of EP was seen as being focused on bringing forward sites in relevant areas, enabling use of public sector land in priority areas for appropriate uses such as housing, enabling land acquisitions or other support for existing organizations, and applying Compulsory Purchase Orders to allow development to take place in priority areas.

Anti-poverty strategies

In addition to formal economic regeneration programmes, many local authorities in the 1990s developed anti-poverty strategies that were directed at the labour market, small business, community enterprises, property development, inward investment and welfare rights, with others focusing on neighbourhood projects. Such projects were supported at national level after 1997 by the Local Government Anti-Poverty Unit which provided research and policy development functions. In terms of approach, many local anti-poverty strategies were geared to a community development orientation. Such initiatives therefore aimed to provide skilled community workers to support community groups. However, there were often tensions within such approaches, since local authorities had to balance the need to promote employability with the need to avoid the labelling or stigmatization of particular areas suffering from disadvantage. Moreover, a wider issue arose from the danger of fragmentation where any type of local anti-poverty strategy was not effectively integrated with government urban programmes.

Policy after 1997: 'New Labour' innovations

Changes to SRB

The newly-elected New Labour Government in 1997 introduced a series of innovations to urban regeneration policy. For instance, because of the problems associated with the Single Regeneration Budget, the government in 1997 introduced a change in the orientation of the SRB. In particular, the guidance provided by DETR in 1997 for Round Four of the SRB indicated that proposals should contribute directly to reducing the multiple causes of social and economic decline. It also suggested that a new emphasis was to be given to the most deprived areas within the SRB initiative. The guidance also indicated that schemes were to clearly integrate with existing strategies, and a more collaborative approach was encouraged. These aspects would seem to underline the lessons from previous initiatives in the 1980s; in particular, it was intended that bids would operate so as to encourage integration of policy at the local level, with so-called 'cross-cutting' policy development, and the involvement of local communities to a greater extent. As a means of facilitating such innovation, the government introduced longer lead-in times for projects, more time for consultation, and more resources for capacity-building. Specifically, all proposed regeneration partnerships were required to make clear how they would link different policy areas within the project proposal; how the strategy fitted other local and regional strategies, and how proposals would integrate with other area-based strategies (Hill 2000).

In addition, the guidance suggested that proposals should seek to contribute to new government policy aims relating to employment, childcare, crime, housing, public health and sustainable development. This new, refocused SRB therefore showed how the problems of the worst neighbourhoods were to be tackled, with the 50 most deprived local authorities being targeted for funding. The new SRB was, together with the New Deal for Communities, to form the core of the new approach to urban regeneration of the government. The key, it was assumed, was targeting, and the SRB was reformed so as to concentrate 80% of the new resources (which comprised around £700 million over a three-year period from 1998) in areas of greatest need (Hill 2000). These broad aims were largely repeated in the bidding guidance for SRB Round Five, which listed the priorities as education and employment skills; addressing social exclusion; sustainable regeneration; local economic growth; and tackling crime and drug abuse. The guidance also stressed the need for the involvement of local communities in the development of SRB bids.

However, in March 2001 it was announced that there would be no further rounds of the SRB, and, instead, Regional Development Agencies were to be given greater flexibility in how they were able to use resources allocated to them via so-called 'Single Pot' funding (the amalgamation of eleven separate funding streams).

'Area-based' initiatives

The New Labour government also launched a series of 'area-based initiatives', targeted on areas and communities where there was a need for priority action. Such initiatives also aimed to support new, cross-cutting approaches involving the concerns of more than one department, and they promoted local partnership as well as community involvement. In addition, they encouraged greater flexibility and responsiveness in the operation of public spending programmes (Department of the Environment, Transport and the Regions 1999).

Typically, such 'area-based' initiatives comprised designated geographical 'zones' with a specific sectoral emphasis. For instance, Education Action Zones, established by the Department of Education and Employment, aimed to encourage innovative and flexible approaches to raising levels of attainment and overcoming barriers to education in areas with high levels of disadvantage. Similarly, Health Action Zones aimed to bring together a wide range of organizations so as to develop and implement a locally-agreed strategy for improving the health of local people. In addition, Employment Zones (with one in Glasgow but the remaining 17 in England and Wales) aimed to improve their employability of the long-term unemployed by means of an implementation programme delivered by local partnerships. Benefits, employment and career services were co-ordinated within such zones, with attention being focused on the over-25s, the disabled, the low-skilled, lone parents, ex-offenders and those made redundant. Other 'area-based initiatives' comprised: the Pioneer Community Legal Service Partnerships; the Crime Reduction Programme (Burglary Initiative and Targeted Policing Initiatives); the Local Government Association New Commitment to Regeneration; New Start; the (revised) Single Regeneration Budget; Sure Start; and the New Deal for Communities (Department of the Environment, Transport and the Regions 1999).

Amongst these 'area-based' initiatives, the New Deal for Communities (NDC) was particularly significant in funding terms since it was comparable in importance to the New SRB. The New Deal for Communities was aimed at offering help to the most disadvantaged communities, by bringing together housing and regeneration programmes and enhancing opportunities for employment and economic development. Initially, 17 'pathfinder' authorities were selected for NDC, and these authorities had to identify a neighbourhood of between 1,000 and 4,000 households on which to target the available resources. A sum of £800 million was initially allocated to the NDC initiative for a three-year period. Significantly, the government specified that the projects in the areas targeted were to be led by community and voluntary groups rather than the local authorities themselves. This was intended to boost the role of the voluntary sector in these local areas, and the local authorities' duty was to develop an overall strategy that would ensure that core issues of unemployment, education, health and poverty were addressed (Hill 2000).

The perceived need for such 'area-based' initiatives as a central component of a set of initiatives aimed at addressing poverty was underpinned by evidence of increasing concentration of poverty in England in the late 1990s. This arose from the 1998 Index of Local Deprivation, which led to the selection of the 44 local authority districts identified as having the most extreme problems. The increased polarisation in town and cities seemed clear from this evidence, with the poorest becoming increasingly concentrated in areas of greatest need. Indeed, the 44 districts selected contained 85% of England's most deprived wards (Hill 2000).

Employment initiatives

It is also important to note other innovations established by the post-1997 New Labour governments that link with area-based initiatives but that are not themselves area-based. Such initiatives were more 'thematic' in nature, often geared to employment. For instance, the New Deal welfare to work programme aimed to bring young people into employment. This programme, which was funded initially from the so-called 'windfall tax' on the utilities, involved provisions designed to bring – some might say coerce – young people into paid employment or at least job placements.

A variant of the New Deal – the New Deal for lone parents – involved the application of increased allowances for childcare costs and other provisions designed to get lone parents back into work. Funding for all the New Deal schemes was committed until 2002. However, it is significant that many local authorities objected to the nature of the New Deal initiatives. In particular, they expressed concern over the quality and the number of permanent jobs that would be created. They were also unhappy with the prominent role the private sector was assumed to play in the New Deal programme. This is because in many areas subject to high rates of unemployment it seemed clear that the capacity of the private sector to produce the required number of jobs was severely limited. In addition, other related 'thematic' innovations introduced by the New Labour government included parallel changes to the tax and credit systems through the Working Family Tax Credit, which aimed to give greater incentives for work. It is interesting in this context to note that all these employment-related initiatives would seem to have their origins – in terms of concepts and inspiration – in the USA. In this respect, they have much in common with many of the other urban policy initiatives introduced since 1968 (Hill 2000).

The Social Exclusion Unit

In conjunction with the above innovations, the government established the Social Exclusion Unit (SEU). This was located in the Cabinet Office and it reported direct to the Prime Minister. Its objective was to enable the development of integrated solutions to problems that were perceived to cover the legitimate areas of a range of departments: so-called 'joined-up' solutions. Hence the Unit was intended to work across what had previously been problematic functional and departmental boundaries, reflecting the so-called 'silo' mentality whereby areas of interest and concern were protected to the detriment of overall integration and co-ordination of policy. One of the early tasks of the SEU was to address the worsening condition of the most disadvantaged council housing estates in England, largely as a result of the kind of evidence of increasing poverty concentration referred to above.

The SEU produced a report entitled *Bringing Britain Together – A National Strategy for Neighbourhood Renewal* in 1998 (Social Exclusion Unit 1998). This Report directly addressed the problem of the worsening council estates though it did not attempt to offer a single definition of a 'poor neighbourhood'. Instead, it referred to areas with complex problems concerning poverty, unemployment, poor health and high levels of crime. It did however refer to the 1998 Index of Local Deprivation in order to justify the areas chosen. It addressed the issue of the failure of previous policy initiatives that had not attacked the causes of urban decline, and suggested that policy was too fragmented, with significant gaps. Consequently, it proposed that in the future policy should be geared directly to addressing the causes of urban decline in terms of poverty and social problems, and so it emphasized the need to address issues of unemployment, homelessness, crime and drugs, young people, health, ethnic minorities and access to services. In terms of policy approach, it indicated the need for integrated strategies that sought to improve employment prospects in particular, to provide support for those with particular problems, and to improve community protection (Hill 2000).

What is perhaps most significant about the SEU as an innovation is its perceived importance within the wider institutional framework of government – essentially the fact that it reported direct to the Prime Minister. This highlights the priority given by the government to addressing the issues of concentrated disadvantage, and the government's emphasis on social exclusion, rather than narrower physical or economic aspects of regeneration, is clearly reflected in the SEU. However, it may be argued that the SEU did not effectively overcome the problems of fragmentation that arose as a consequence of the plethora of individual initiatives introduced after 1997. In 2006, the SEU was superseded by the Social Exclusion Taskforce, intended to concentrate on the people most at risk from persistent social exclusion, including hard-to-reach groups (Cabinet Office 2006).

Neighbourhood Renewal

Following the SEU's 2001 report *A New Commitment to Neighbourhood Renewal* (Social Exclusion Unit 2001) there was a new emphasis on addressing mainstream services, targeted to the 88 most disadvantaged areas, with a strategic and co-ordinated approach. These areas were required to establish Local Strategic Partnerships (LSPs) which comprised coalitions of public, private, community and voluntary sectors. These LSPs were required in turn to produce neighbourhood renewal strategies to show how actions were to be targeted and co-ordinated, and these strategies were to be funded by a new Neighbourhood Renewal Fund (NRF). The report also required the setting-up of 'floor targets', or minimum standards below which no neighbourhood should fall, and it encouraged neighbourhood management, with neighbourhood managers being responsible – via a local board – to the LSP. Community involvement was also prioritised within such activities, and separate funding was made available via the Community Empowerment Fund (CEF), to support such activity.

Urban Task Force

The Urban Task Force, chaired by Lord Rogers of Riverside, produced a report entitled *Towards an Urban Renaissance* (Urban Task Force 1999), which set out a series of recommendations for all aspects of urban policy in England. However, the Task Force Report was criticized by some for an over-emphasis on physical issues. Nevertheless, several of its recommendations, including the creation of urban regeneration companies, were carried forward into the subsequent Urban White Paper *Our Towns and Cities: the Future. Delivering an Urban Renaissance* (Department of the Environment, Transport and the Regions 2000). In fact, a great number of the Task Force's recommendations were not carried forward into the White Paper, and in 2005 the Task Force Members produced a further document – *Towards a Strong Urban Renaissance* (Urban Task Force 2005) – which indicated where there were remaining gaps). Specifically, this Report acknowledged the achievements of the post-1997 Labour governments in terms of urban regeneration, including the attraction of many people back to city centres to live. However, it also suggested that the quality of the built environment was still not at the core of the agendas of key agents such as Urban Development Corporations, Urban Regeneration Companies and Regional Development Agencies. In addition, it indicated the need for integration of issues such as transport more clearly within regeneration, suggesting that the Department for Transport was not part of the government's regeneration agenda.

The Report recommended a series of measures to take forward the need for greater awareness and application of good urban design, as well as more effective public participation processes. In terms of social wellbeing, it suggested that progress had been made for instance in terms of crime prevention measures, and by housing associations in widening housing choice. Nevertheless, it indicated that there were still major problems of housing affordability, social polarization and concentrations of disadvantage within urban areas, and it therefore suggested further action to ensure greater social and economic mixing within communities. In terms of institutions and initiatives, the report highlighted the problem that 'a plethora of overlapping, but differently funded and monitored, area-based regeneration bodies has reduced the delivery effectiveness of public sector regeneration schemes. This has been exacerbated by the disconnection of regeneration expenditure between Government Regional Offices, Regional Development Agencies and English Partnerships and the huge number of new ineffective partnerships at local and sub-regional levels, particularly in areas like the Thames Gateway' (Urban Task Force 2005, 15). In addition, it called for more incentives for re-use of brownfield sites, greater involvement in site assembly on the part of local authorities, a revised remit for RDAs to include regeneration to an equivalent degree to economic development, the empowerment of city governments to raise funding for regeneration, and a move towards one area-based delivery body per regeneration area.

It is significant, however, that Professor Sir Peter Hall indicates in a footnote to the Report that, whilst he is agreement with the great majority of the report, he is unable to support certain conclusions in relation to use of brownfield land. This is because, he suggests, the Report's priorities for greater re-use of brownfield land, and for the raising of minimum residential densities, would deepen existing problems in relation to housing provision. This reflects the controversy that continues to surround aspects of regeneration concerned with provision of new and affordable housing, and while such debates are of particular importance for the south east of England, where growth pressures (and limitations of available land) are most intense, aspects of the debate are also of clear relevance in a wider context.

Urban Regeneration Companies

The first Urban Task Force Report (Urban Task Force 1999) also recommended the creation of Urban Regeneration Companies to stimulate new investment into areas of economic decline, and to co-ordinate plans for the regeneration of such areas. Consequently, English Partnerships developed the URC model, and three pilot projects were launched in Liverpool, East Manchester and Sheffield. Essentially, URCs are independent companies established by the relevant local authority and Regional Development Agency. They unite public and private sector partners, and they work alongside English Partnerships and other local stakeholders including employers, amenity groups and community representatives. Their aim essentially is to involve the private sector in a sustainable regeneration strategy in the context of a wider strategic framework or masterplan taking into account the problems and opportunities of the wider area. Their focus is on the unlocking of significant but latent development opportunities, and thereby realizing a collective vision for the future of an area.

Following the success of the pilot URCs, the *Urban White Paper* in 2000 (Department of the Environment, Transport and the Regions 2000) proposed the creation of a further 12 URCs, and in 2006 URCs had been established in 20 locations, with several other areas developing proposals for an URC. Early evaluation of the URCs in England and Wales indicated that they were able to increase the confidence of the private sector to invest in defined areas, by promoting areas as well-managed, ensuring the commitment to joint working of key decision makers, and bringing about early actions as the basis for growth. Significantly, however, the 2005 Urban Task Force report (Urban Task Force 2005) indicated that the Urban Regeneration Companies that had been established lacked the necessary powers to fulfil their role.

Sustainable communities

The Government's sustainable communities programme is also relevant here. As set out in the ODPM's document *Sustainable communities: building for the future* (Office of the Deputy Prime Minister 2003), this programme sets out a long-term vision to address problems of housing shortage, affordability and abandonment. An important part of the programme is the identification of opportunities for sustainable housing development in the south-east of England. In addition, however, it involves action to address problems of poor housing quality, abandonment and homelessness in areas subject to urban decline, many in the north of England.

Conclusions

Overall, the trajectory of urban regeneration policy development since the 1960s would seem to accord with the typology of Giddens as set out in Chapter 2, particularly in terms of the implications of the neo-liberal period of the 1980s, which led to the adoption of mechanisms such as UDCs and Enterprise Zones, together with the introduction of competitive bidding for urban regeneration funding. Post-1997 policy, it may be argued, can be seen to reflect many aspects of the 'Third Way' notion, including a greater emphasis on community involvement for instance, but with a clear continuation of many aspects of the neo-liberal approach.

In general terms, however, there would seem to have been a progression in the development of policy since the 1960s in many respects, for instance in terms of increasing recognition of the need for inclusive and integrative policy. Hence there is evidence of an increasing recognition of the need for 'holistic' and integrative policy approaches that seek to address social, economic and environmental objectives, though the realization of this objective has been more problematic. In addition, the need for inclusivity of policy development and application has increasingly been acknowledged, with inclusive partnership now forming a central element of most regeneration initiatives and funding mechanisms. Certainly, the profile of regeneration activity remains high in many areas of government, particularly at the local level.

However, it may be suggested that an over-emphasis on structures rather than processes of partnership, as set out in Chapter 2, remains, with a continued plethora of partnership structures and institutions in many areas illustrating a somewhat 'crowded platform' for regeneration activity. Moreover, there remain persistent unresolved tensions and issues within urban regeneration policy. Notably, many argue that such policy remains marginal within the overall spectrum of public spending programmes, with a parallel failure to create meaningful links between area-based programmes and mainstream spending. In addition, it may be argued that urban regeneration policy is still not adequately geared to local needs. These issues and tensions are clearly embedded in the overall approach and ethos of contemporary governance. As such, they are closely linked to parallel debates and arguments in relation to spatial planning and development, and the need (proposed by government) for this area of activity to be streamlined and 'modernized' so as to facilitate economic growth and restructuring. More specifically, contemporary issues and tensions with respect to urban regeneration policy in England may be characterized as follows.

Policy aims

A persistent problem with urban regeneration policy in England may be seen to arise from the lack of a coherent definition of what it seeks to achieve. This was a particular problem in the 1980s, when there was confusion over whether policy was intended to improve the competitiveness of the economy as a whole, or to spread the benefits of economic development, though the former seems to have been dominant during this period. However, this may be a false dichotomy; as Robson suggests, 'long-term economic effectiveness can only fundamentally be achieved if the whole of the population derive some benefit from it' (1994, 176). Nevertheless, leakage of benefits through labour markets implies the need for policy to be directly addressed to those most in need. While this issue would seem to have been addressed to some extent by the new Labour governments after 1997, there would seem to have been a failure to effectively articulate the aims of urban regeneration in such a way that they could be linked with mainstream policy objectives.

Spatial targeting

An important issue that underlies much of urban regeneration policy since 1968 in England is that of geographical targeting. In terms of employment, for instance, Robson (1988) suggests that jobs created in inner city areas have not necessarily been of benefit to these areas because of the open nature of labour markets, indicating that spatial targeting is ineffective. Moreover, Parkinson (1988) suggests that spatially-targeted, area-based initiatives tend to displace problems between neighbourhoods, create a long-term dependency culture, fail to address the problems of individuals outside targeted areas, and fail to tackle the causes of urban decline which lie outside disadvantaged areas (Parkinson 1998). Nevertheless, there seems to be a role for such area-based approaches within urban regeneration, since they can help to compensate for local market failure, increase community capacity, linking local policy to city-wide regeneration strategies, and encourage policy integration at the local level.

Indeed, a 2006 report by the Joseph Rowntree Foundation indicates that there is some evidence that area-based renewal has contributed to the 'turning-round' of the most disadvantaged housing estates in England. However, the Report also suggests that this may have been in large part the result of broader policy and economic and social change, and in any event many residents of such areas would still appear to be significantly disadvantaged by living in such areas. So further and sustained area-based action, it suggests, is needed to ensure the consolidation of improvements. In addition, the Report acknowledges that improvements have been largely physical, and changes in social composition and levels of disadvantage may have been largely the result of people being enabled to move out of the area, again reflecting the limitations of an area-based approach (Tunstall and Coulter 2006).

Co-ordination of policy

In addition, the danger remains in the English context of too many separate initiatives, with a lack of attention given to the organizational infrastructure necessary to sustain collaborative methods of working (Clarence and Painter 1998). Certainly, the bewildering number of regeneration initiatives created by the new Labour governments after 1997 led to inefficiencies and confusion at the local level, in spite of the aspiration for integration and co-ordination of urban regeneration policy development and application. While the creation of the Social Exclusion Unit represented a creditable approach to promoting a more 'holistic' and preventative approach to urban regeneration, it may be argued that it did not clearly address the fundamental and entrenched issues of inter-departmental rivalry and a fragmented institutional context. This issue is compounded by the tension between the need to target resources to defined groups and the need to target defined areas. In order for both these approaches to be effectively integrated, it is necessary for linkage to be made with mainstream local services; however, the large number of initiatives introduced after 1997 – as indicated above – would seem to have made this more difficult (Hill 2000).

Institutional infrastructure

Clearly, contextual factors, particularly those concerning governance (at both local and national levels) must be taken into account in considering how regimes for urban regeneration can be made more effective. Indeed, it is significant that England's Urban Task Force report (Urban Task Force 1999) highlights the need for fundamental changes in local and national modes of governance in the UK. In particular, it suggests that central government departments, driven by service-based policies, were far too powerful, and asserts that 'The single biggest priority of the Government in developing and implementing its regeneration policy over the next few years is to break down central government departmentalism' (1999, 138). Partly as a result of such problems, the Urban Task Force Report suggests the need for 'strong strategic local government, which can provide long term vision and which can consider in a holistic way all the major needs and opportunities of a town or city, with the engagement of its people' (1999, 44). This is at odds with much practice at central government level, since UK governments have in recent decades managed to combine a decentralising rhetoric with a centralising instinct (Clarence and Painter 1998). Moreover, it may be argued that, in spite of widespread re-structuring, local government continues to suffer from problems of departmentalism that inhibit local collaboration and joint working, though the culture of protectionism that arguably existed in the 1980s and early 1990s, as a response to restrictions imposed by central government, would seem to have gone.

The 2006 Local Government White Paper (Department for Communities and Local Government 2006) reflects such debates on localism and governance for instance by making it easier for communities to opt for directly-elected mayors or executives, and thereby allowing stronger local authority leadership. It also reflects the need for strategic policy development by encouraging co-operation and partnership between local authority areas, for instance by 'multi-area agreements' which allow the pooling of funding across conurbations for strategic functions such as regeneration. This may allow a greater role for strategic regeneration policy. Nevertheless, it may be argued that the White Paper's provisions do not allow for the strength of local governance as suggested for instance in the 1999 Urban Task Force report, and they do not specifically address the issue of city-region governance.

Chapter 4

Urban Regeneration Policy in Scotland

Introduction

This Chapter outlines the evolution of urban regeneration policy in Scotland and the main points of difference with the evolution of such policy in England, with reference to the important institutional differences in context and the differing spectrum of actors involved in the development and application of policy.

Essentially, the shift in priority in England in 1977 from new towns to inner city areas was paralleled in Scotland in 1976 by a shift in resources from a planned new town at Stonehouse to the Glasgow Eastern Area Renewal (GEAR) project in Glasgow (Keating and Boyle 1986; Lawless 1989). Moreover, Turok (1987) identifies three phases of urban policy which can be seen to apply to both England and Scotland, namely a social orientation in policy until the mid-1970s, a concentration on economic factors and employment in the late 1970s, and an increasing reliance on the private sector after 1979. However, the evolution of many urban policy mechanisms was somewhat different in England and Scotland (McCrone 1991), as indicated in more detail later in this Chapter, and major differences persist in this respect. Indeed, Turok (1987) acknowledges the significant role played by the Scottish Development Agency, and the different nature of the Scottish Urban Programme, with its greater emphasis on social programmes, at least until the mid-1980s, partly because of the importance of community businesses in Scotland.

Scottish Development Agency

The significant nature of many Scottish urban regeneration policy initiatives since the 1960s would seem to have arisen from the legacy of municipal intervention as well as the important emphasis on public sector intervention as implemented by the SDA (Boyle 1988). The Urban Renewal Unit, set up in the Scottish Office in 1975, is important in this respect, since it was intended to 'formulate broad urban policy based on co-ordinated, controlled positive discrimination for selected urban renewal schemes, encompassing both socio-economic and physical change' (Keating and Boyle 1986, 6). In fact, the Urban Renewal Unit considered the possibility of establishing a new agency with powers similar to the Urban Development Corporations (UDCs) in England and Wales. However, this option was not pursued and the SDA was set up instead, with the aim of enabling economic regeneration in partnership with the existing agencies (Keating 1988).

The Glasgow Eastern Area Renewal (GEAR) scheme, co-ordinated by the SDA, may be seen as the organization's 'flagship' project. Nevertheless, some observers suggested that it provided evidence of the agency maintaining an appearance of partnership whilst usurping the role of local government (Atkinson and Moon 1994; Robson 1988), and other SDA initiatives co-ordinated by the SDA would seem to have achieved more effective partnership.

In fact, several separate phases of policy application can be distinguished within the overall approach of the SDA. In this context, the GEAR programme may be seen as the conclusion of a period of traditional, comprehensive urban renewal. Subsequent integrated 'area projects' may be seen as local intervention using partnership methods, while later projects based on a 'self-help' model may be seen as reflecting an increasing reliance on enterprise and business development (Boyle 1988; Lawless 1989; Atkinson and Moon 1994). The particular technique of targeting policy to small, concentrated areas was promoted after 1981, when the SDA's area programme development division was set up to develop relevant policies, together with an area development directorate intended to implement such policies (Keating and Boyle 1986; Keating 1988). The SDA decided to spend 60% of targetable funds on the 'area projects', and it initially selected areas considered to exhibit potential, rather than the worst areas of need, because of its basic aim of creating internationally-competitive industry. However, this policy was contentious, and it was subsequently decided to select areas where there was the biggest discrepancy between potential and performance (Keating 1988).

The later focus by the SDA on business development and private sector 'opportunities' (Keating 1988; Turok 1987) followed the prevailing trend in England, and led to criticism that commercial potential became the most important criterion for the selection of projects. This may clearly be seen in the case of Glasgow, where resources were concentrated in the city centre at the expense of the peripheral estates (Keating 1988; Boyle 1988). Such examples clearly illustrate a downgrading of the SDA's original social objectives of urban regeneration (Keating and Boyle 1986).

These aspects of urban regeneration policy illustrate some important differences in the Scottish context as compared with England. However, several elements of the Scottish approach have been similar to those in England. For instance, the Local Enterprise Grant for Urban Projects (LEG-UP), administered by the SDA (and later, the local enterprise companies) to encourage private investment in urban areas, was similar in operation to the English Urban Development Grant (UDG). In addition, the aims of the 'Action for Cities' initiative in England were largely matched in Scotland by the *New Life for Urban Scotland* White Paper (Scottish Office 1988) which encouraged a locally-co-ordinated approach to urban regeneration. In the case of both approaches, the emphasis was on a more effective use of existing resources and an increasing contribution from the private sector, rather than on additional funding.

However, the Scottish White Paper also involved a significant shift in concern away from the problems of inner urban areas and towards those of the outer peripheral housing estates, as indicated in Chapter 7. These areas had been developed in response to serious housing shortages, though a shortage of development funding meant that development costs per unit had been minimized, with the consequence of early obsolescence (McCrone 1991). Four Housing Partnerships were subsequently selected for comprehensive regeneration, and these initiatives also highlighted the role of the local community in regeneration. Indeed, the later 1990 Scottish Office report *Urban Scotland into the 90s* gave a similar level of prominence to the local community as to the private sector (Atkinson and Moon 1994), as well as highlighting the need for greater linkage of physical and economic regeneration (Scottish Office 1990).

As in England, the Scottish Urban Programme was extremely important as an urban regeneration initiative. However, it was criticized as being too small, and somewhat misdirected (Keating and Boyle 1986). Moreover, Gillett (1983) suggests that the Urban Programme was 'not a policy designed for Scotland, and changes in it have to an extent followed the line taken in England' (118). This is supported by the *Programme for Partnership* policy statement on urban regeneration published by the Scottish Office in 1995 (Scottish Office 1995), which indicated that future funding would be allocated as part of strategies and programmes submitted by local authorities or other regeneration partners, rather than for individual projects.

This marked an important point in the development of partnership approaches, since it involved the use of limited resources more strategically to achieve maximum impact in areas of greatest deprivation (Bailey *et al* 1995). This reflected the shift to a more focused approach to UP funding in England in the late 1980s, which involved an emphasis on incorporation of schemes within a strategic approach to specific priority areas within UP authorities. While the English Urban Programme was subsequently phased out in favour of the more integrated, competitive approach of the SRB, the new emphasis on a strategic approach to regeneration remained.

Priority Partnership Areas

The Priority Partnership Area (PPA) initiative emphasized the designation of communities for practical support for specific projects within a city-wide regeneration strategy. This distinctive combination of policy elements reflected the historical development of policy for urban growth and regeneration in Scotland. The Priority Partnership Area concept, initiated in 1996, was an outcome of the corporatist tradition in Scotland. It drew together central and local government, the private sector and a host of other partners which came together in a strategic framework for urban regeneration focused on specific geographical neighbourhoods. Furthermore, the PPAs were devised on the basis of community involvement in the design and execution of projects, and they operated through the medium of partnership. The emphasis was on outputs, namely tangible improvements in the conditions of the most disadvantaged communities in the areas designated for support, which were measured by key indicators of economic and social change.

The PPA policy initiative in Scotland encouraged the formation of city-wide partnerships, an associated city-wide urban regeneration strategy, and more specific proposals intended to address the problems of social and economic disadvantage in defined neighbourhoods. It was developed essentially as an extension of the principles laid down in the Scottish Housing Partnerships, which themselves had concentrated on the 'holistic' regeneration of housing areas. These had been restricted in scope to the four peripheral housing areas designated by the (then) Scottish Office. However, the main principles of co-operation, integration and partnership, within a strategic regeneration context, and the emphasis on the involvement of local communities, were felt to be applicable to broader urban regeneration funding. The Scottish Office therefore announced a new competition for regeneration funding which would use these features as the basis for selection of funding schemes. The results of this competition were announced in November 1996, and PPAs were designated in the following local authorities: Aberdeen, Dundee, Edinburgh, Glasgow, Easterhouse, Inverclyde, North Lanarkshire, Renfrewshire, South Ayrshire and West Dunbartonshire.

The PPAs involved the development of local partnerships which could be developed from existing arrangements, such as the Dundee Partnership, which comprised amongst others the City Council, Scottish Enterprise Tayside and Scottish Homes. However, many PPA partnership groupings were specially formed for the purpose, arguably representing 'impermanent alliances' formed for the immediate purpose of levering out comprehensively allocated public sector resources. Whatever the origins of the PPA arrangements, a key theme which ran through them was the need for a city-wide urban regeneration strategy. Within such a strategy, specific proposals were put forward for designated PPAs, based on the most disadvantaged neighbourhoods in the city.

The key indicators that characterized such communities as disadvantaged included youth and long term unemployment, low income households, uptake of benefits and education support grants, levels of educational attainment, crime, mortality rates and other health indicators, which were all significantly worse than those for the city as a whole. In addition to the PPAs, the Scottish Office devised the notion of Regeneration Programmes, for areas which were smaller in scale than PPAs but which nevertheless exhibited similar concentrations of disadvantage. Essentially, the new designations represented new ways of channelling Urban Programme funding to specific areas, though the Urban Programme remained the principal source of funds for regeneration initiatives in Scotland at the time.

Social Inclusion Partnerships (SIPs)

The PPA initiative was superseded by a new measure – the Social Inclusion Partnerships (SIPs). This represented the next stage in the development of policy for urban regeneration, namely the emergence of mechanisms for addressing the dynamics of social disadvantage through preventing social exclusion and securing social inclusion. Essentially, the SIP initiative was intended to extend the principles embedded in the earlier PPA measure to directly address the dynamics of social inclusion and social exclusion. The introduction of SIPs was announced in 1998 with the intention that the 12 existing PPAs would evolve into SIPs and that additional new SIPs would be set up. It is clear that SIPs represented a new phase in Scottish urban regeneration policy; while the measure was drawn directly from the government's approach to social exclusion, it reflected a specifically Scottish theme. For example, in introducing the measure, the Secretary of State for Scotland argued that Scottish circumstances differed from England in that those suffering exclusion in Scotland were disproportionately concentrated in specific communities. Social Inclusion Partnerships built on established arrangements and experience, for instance in relation to the PPAs, but with a particular emphasis on seeking to prevent young people, in particular, from being excluded in participation in the economic and social mainstream. The key characteristics of SIPs were:

- a focus on the most needy members of society to reduce turnover rates, stabilise the community and attract people onto the estate;
- the co-ordination and filling in of gaps between existing programmes in order to promote inclusion; and
- an attempt to prevent people becoming socially excluded.

It was therefore intended that the SIPs would focus to a greater extent on promoting inclusion and preventing social exclusion from developing, particularly in terms of the needs of children and young people. In this sense, the SIPs initiative represented to a large extent a continuation of the established approach, but with a key innovation in its emphasis on addressing the perceived dynamics of exclusion and inclusion. In practice, it was intended that existing PPAs would evolve into SIPs and the new SIPs programme would be opened up to those communities that missed out on the initial PPA designations. It followed that established PPAs had to adjust their arrangements to reflect the expectations of the new emphasis of the SIPs approach. In particular, the new approach included:

- sectoral initiatives in education focused on promoting inclusion for children and young people with an emphasis on 'New Deal Schools' and 'Out of School Learning Projects';
- pilot New Deal Pathfinder projects aimed at strengthening community capacity; and
- pilot Pathfinder projects in Wester Hailes and Easterhouse which were to develop new integrated approaches to service delivery by unitary local authorities.

An achievable partnership approach remained central to the designation process, together with a long term strategy and the involvement of local communities. In this respect the Scottish Office expected that partnerships seeking support would demonstrate convincing strategies, effective partnership, and ability to deliver. The SIPs initiative also involved an additional innovation, since it introduced a shift away from an exclusive reliance on geographically focused policy. Among the suggested approaches to be adopted in the SIPs framework were sectoral measures which could cut across defined areas of disadvantage and address processes of social exclusion in a city-wide context. In many respects, therefore, the SIPs model built on and extended the ideas enshrined in the earlier PPA approach to urban regeneration.

Community Regeneration Fund

The Social Inclusion Partnerships funding programme, together with the Better Neighbourhood Services Fund (BNSF), were replaced by the Community Regeneration Fund (CRF) in 2004. The Scottish Executive established this fund, to run from 2005-08, in order to bring improvements to Scotland's most deprived areas, and to help people escape poverty. More specifically, the aim is to promote the community regeneration of the most deprived neighbourhoods by means of clear improvement (by 2008) in employability, education, health, access to local services, and quality of the local environment. The Fund is therefore intended to enable people living in the most disadvantaged neighbourhoods to take advantage of job opportunities and to improve their quality of life. In addition, in conjunction with other programmes, it seeks to lift such people out of poverty, improve the confidence and skills of young people, reduce the vulnerability of low income families to financial exclusion and debt, increase the rate of health improvement for those in the most deprived communities, and improve their access to services.

The CRF is targeted on the 15% most disadvantaged communities as identified in the Scottish Index of Multiple Deprivation. It comprises £318 million, and its application builds on the experience of the SIP programme in particular. The CRF was managed by Communities Scotland, the Scottish Executive agency with responsibility for housing and regeneration, which aims to promote social justice and tackle exclusion by means of the delivery of sustainable community regeneration. The various Community Planning Partnerships in Scotland produce three-year Regeneration Outcome Agreements (ROAs), which set out how they intend to use the CRF, in conjunction with their own resources, to deliver specific regeneration outcomes.

Cities Review and Cities Growth Fund

The Review of Scotland's Cities, the so-called 'Cities Review', in 2002 (Scottish Executive 2002b), was essentially the Scottish version of the Urban Task Force report in England (Urban Task Force 1999). As such, the Cities Review sought to review the prospects for the economic, environmental and social development of the five Scottish cities (Aberdeen, Glasgow, Edinburgh, Dundee and Inverness), and to identify Executive policies which would improve these prospects (Scottish Executive 2002b). The remit of the Review was to consider economic, environmental and social developments in the cities concerned, and it did not therefore seek to look in detail as aspects such as health. The Report of the Cities Review was not a statement of Scottish Executive policy, but it was accompanied by a policy response from the Executive, entitled *Building Better Cities: Delivering Growth and Opportunities* (Scottish Executive 2003c).

Following the Cities Review for Scotland, the Cities Growth Fund was created to support key growth projects in Scotland's cities, including Stirling. This Fund invested £90 million from 2002 to 2005, and a further £82 million was allocated for 2006 to 2008 (Scottish Executive 2003c).

Community planning and Community Planning Partnerships

Following its election to office in 1997, the Labour Government asserted the importance of enhancing the leadership and enabling roles of local government. This brought together ideas of leadership and well-being, and improved participation from diverse local communities (Kitchen 1999). In England and Wales, this led to the Local Government Act 2000, which introduced a power of community well-being, and the concept of community strategies. In Scotland, the same principles and concerns about appropriate local governance resulted in the legislative provision for community planning. As a process, community planning also includes measures for regeneration.

The Local Government in Scotland Act (2003) provides the statutory basis for community planning, alongside related aspects of local governance, including arrangements for 'best value' and the provision of a power to advance well-being in communities. The main aims of community planning are to ensure that communities are genuinely involved in decisions made regarding public services, linked to a commitment from organizations to work together to provide better services (Scottish Executive 2003a). Essentially this was the foundation of the modernization of arrangements for regional and local governance in Scotland. The Partnership Agreement between Labour and the Liberal Democrats, which set out the programme for the second session of the Scottish Parliament, for example, envisaged a Scotland where enterprise could flourish and where there was opportunity for all (Scottish Executive 2003b). To this end, the Executive encouraged the delivery of local services of the highest possible quality to promote the empowerment and inclusion of local communities.

Community planning in Scotland is bound up with the work of the Scottish Parliament, its Executive, and the reconfiguration of relationships with local authorities (Alexander 1997). Community planning was advocated, therefore, as a means of re-asserting the role of local authorities at a time when new relationships, policy agendas and national priorities were being developed in Scotland (Sinclair 1997). In other words, community planning may be seen as representing an attempt to provide a strategic framework for the activities of the multifarious institutions engaged in local governance, community capacity building and regeneration.

The specific aims of community planning are as follows:

- to improve the service provided by Councils and their public sector partners to the public through closer, more co-ordinated working;
- to provide a process through which Councils and their public sector partners, in consultation with the voluntary and private sector, and the community, can agree both a strategic vision for the area and the action which each of the partners will take in pursuit of that vision; and
- to help Councils and their public sector partners to collectively identify the needs and views of individuals and communities and to assess how they can best be delivered and addressed.
(Community Planning Working Group 1998).

The Community Planning Working Group (1998) argued that there was a need to establish a more systematic approach to ensure greater co-ordination between the processes in place for service delivery, and the communities for which they were intended. It concluded that 'there is at present a lack of structured overview across the various agencies about how they collectively could best promote the well being of their communities. The kaleidoscope of local and subject-specific plans and partnerships was not related to a consistent attempt to develop a shared strategic vision for an area and a statement of common purpose in pursuit of that vision. Neither was it always clear who had the lead in developing this shared strategic vision' (The Community Planning Working Group 1998, 3).

For an individual Council area, the purpose of the community planning process would be to present an informed view of the challenges and opportunities facing the geographical communities and the different communities of interest. A community plan was envisaged as applying for between five and ten years, and to be subject to annual review, with clear statements of progress in relation to an agreed agenda for action. It would involve full consultation with communities, voluntary organizations and the private sector, though it would be driven from within the public sector. The community planning process was also envisaged as working at more local community levels of interest.

The concept of community planning involved an outcome (the Plan), as well as a process of negotiation whereby the different interests and policy positions of all the relevant bodies were drawn together into a common agenda for action. In particular, it was intended that community plans would be prepared by local authorities, enabling them to demonstrate greater leadership. This would involve a process within which all local agencies would submit their annual plans to the Council. On the basis of this consultation, the Council would then produce a community plan which 'would incorporate not only the Council's own proposals, including a statement of the standards and quality of service it would provide to the local community, but also the plans of the local appointed bodies and how these plans would contribute to the overall well-being of the community rather than, as at present, being developed and published in isolation' (Sinclair 1997, 17).

As a consequence of the introduction of community planning, Community Planning Partnerships now bring together a range of public sector providers such as local authorities, the National Health Service, police, fire services and the local enterprise networks, as well as the communities that are served by such bodies. The aim of the Partnerships is to ensure that communities are better served by a process of integrated service planning. There are 32 Community Planning Partnerships, one for each of the local authority areas in Scotland. Such Partnerships set out how they intend to use their allocation of the Community Regeneration Fund, in addition to their own resources, in order to deliver clear and measurable improvement in terms of regeneration outcomes. This is done by means of the three-year 'Regeneration Outcome Agreements', which are assessed by Communities Scotland.

Urban Regeneration Companies

Urban Regeneration Companies (URCs), as indicated in Chapter 3, are special purpose vehicles intended to bring together the public and private sectors so as to drive forward the achievement of complex urban regeneration initiatives. Specifically, an Urban Regeneration Company is a formal partnership of key representatives from the public and private sector who operate at arms length to deliver physical and economic regeneration in specific areas, and it povides a strategic overview of an area so as to guide investment decisions by the public and private sectors so as to further specific outcomes. Urban Regeneration Companies were developed in England, and evaluation of their experience showed that they had significant potential to increase the confidence of and investment from the private sector, since they were seen to bring about well-managed areas which applied clear commitment from key stakeholders to work together to achieve common objectives. The involvement of stakeholders and the effective communication with residents are seen as necessary elements of URCs. It is intended that URCs will lever investment from the private sector of over £400 million in order to support the development of infrastructure and land improvements, new business parks, new community facilities and new housing and environmental improvements.

In Scotland, Pathfinder URCs have been established in Clydebank, Craigmillar in Edinburgh and Raploch in Stirling. These are to operate for between ten and fifteen years, and the budget for these URCs was £20 million for 2004-06, with a further £22 million allocated for 2006-08. In addition, start-up funding was announced in 2006 for two further Pathfinder URCs for Irvine Bay and Riverside Inverclyde. This funding will assist with the business planning and land development, and the Executive will consider future funding for these URCs on the basis of detailed business plans. In addition, the Executive is also working with Glasgow City Council, South Lanarkshire Council and their partners to establish a URC for the Clyde Gateway.

People and place: regeneration policy statement

In 2006, the Scottish Executive published a regeneration policy statement which set out its aspirations for regeneration in Scotland (Scottish Executive 2006a). Crucially, the foreward, by Minister for Communities Malcolm Chisholm, links people, place, partnership and prosperity. It asserts, like other Scottish Executive policy documents, the primacy of growing the economy in a sustainable manner, as well as the importance of cities as key economic engines. It also indicates that successful regeneration, seen as the lasting transformation of places and communities, is key to achieving sustainable economic growth. It suggests that environmental and social objectives are important, as well as economic objectives, and that these should be mutually reinforcing. The role of Community Planning Partnerships is endorsed in this document, since these are seen as providing the lead strategic role in regeneration, through the development and delivery of Regeneration Outcome Agreements (ROAs).

In addition, the document proposes the exploitation of the comparative advantage of the city-regions in Scotland by investing in further new business locations, bringing vacant and under-used land back into productive use, and creating further employment opportunities. Priorities, it suggests, must be framed with regard to both opportunity and need. Specifically, the determination of geographic regeneration priorities is to be determined by consideration of: market opportunity, availability of labour, and the extent to which concentrations of deprivation and vacant and derelict land can link to emerging economic opportunities. Clearly, therefore, the document reflects an assumption that spatial regeneration priorities must not be derived solely from the need to exploit economic opportunities. In this context, reference is also made to the economic development zones that are set out in the National Planning Framework, in particular the need for co-ordinated action for the Clyde Corridor and West Edinburgh, as well as to other significant initiatives in progress, such as the development of Ravenscraig. It is also intended to support major regeneration initiatives with the potential to achieve a regional impact, initially in Inverclyde and across Ayrshire, with the intention that these schemes will involve benefits that will be dispersed throughout the regional economy.

The 'holistic' nature of regeneration is therefore acknowledged, and attention is given in the document to explicitly environmental aspects that link with economic objectives, by means for instance of the use of more effective use of Compulsory Purchase Orders (CPOs) to assist in the process of land remediation and assembly. Moreover, aspects of liveability in terms of environmental quality are highlighted by means of better design and better quality public spaces and amenities. In terms of broader health and quality of life, the importance of sport and cultural facilities is emphasized, as are social aims for instance by means of the creation of mixed and diverse communities, which are seen as likely to be sustainable in the longer term. Tenure mix is important in achieving such aims, and greater choice and diversity in housing is seen as necessary in order to offer opportunities for those moving up the housing ladder.

The document therefore sets the framework for regeneration activity in Scotland, but it acknowledges that: 'our approach must be informed by the knowledge, experience and practice of those involved "at the coal face" of regeneration across Scotland, in the rest of the UK and elsewhere ... Our role at national level is to set high level priorities; to join up investments from different sources; to align our activities more effectively; and to work with others to create the right environment for regeneration. Above all it is our job to remove the barriers to action by others and to ensure that the public sector is alive both to regeneration opportunities and the needs of the private sector' (Scottish Executive 2006a, 48).

Overall, the Scottish Executive estimates in this document that from 2005-08, £2.4 billion will be invested in a variety of regeneration programmes, comprising funding from agencies including Communities Scotland, Scottish Enterprise and Highlands and Islands Enterprise, local government, health boards and similar organizations. This includes £123 million via the Cities Growth Fund aimed at enhancing city competitiveness, and £318 million via the Community Regeneration Fund, aimed at improving the most disadvantaged neighbourhoods (Scottish Executive 2006a).

The policy statement identifies the following specific lessons for future policy application, building on the lessons of regeneration from the rest of the UK in recent decades.

The need to apply more than physical regeneration

The quality of local environments is critical to the overall success of regeneration initiatives, as indicated most clearly by poor quality housing and a lack of local facilities. The document suggests that such problems have been addressed to an extent in England by the application of programmes such as the Neighbourhood Renewal Programme, and in Scotland this has been done via the Better Neighbourhood Services Fund. Moreover, it indicates that programmes such as the Cities Growth Fund and the focus of Scottish Enterprise on creating competitive places have aided the implementation of projects with a physical or environmental improvement element. However, there is, it suggests, a need to ensure that regeneration also achieves clear social and economic benefits for local communities (Scottish Executive 2006a).

Linking needs and opportunities

The document asserts that in England, while programmes such as the Neighbourhood Renewal Fund and New Deal for Communities addressed many issues such as health, education, housing and worklessness, underlying economic issues often remained largely intact, with the resulting persistence of concentrations of unemployment and deprivation. It therefore concludes that more needs to be done in Scottish contexts to ensure that local people are able to take advantage of employment opportunities, for instance by means of the development of work skills and broader aspects of employability, though such issues are being tackled to some extent by the Community Regeneration Fund for instance (Scottish Executive 2006a).

Ensuring integration of policy

While integration of policy remains a barrier to regeneration progress in some respects, the introduction of community planning and the production of Regeneration Outcome Agreements, the document suggests, is beginning to address this issue. It also indicates the need to ensure the integration of local approaches with broader-scale consideration of strategic planning in relation to land use, transport, housing and public services. The need for integration may also, it is suggested, be applied to regeneration funding streams. This was applied in England via the Single Regeneration Budget and the subsequent 'Single Pot', and in Scotland progress would seem to have been made in bringing together funding streams such as the Social Inclusion Partnerships, Better Neighbourhood Services, and Tackling Drugs Misuse programmes, which were brought together to form the Community Regeneration Fund (CRF). Nevertheless, the document indicates the need for greater co-ordination and simplification of the framework for regeneration funding in Scotland, as well as the linkage of other funding streams, such as European funding, more closely to regeneration objectives (Scottish Executive 2006a).

Clear leadership and focus

The document suggests that the (arguably narrow) success of initiatives such as the English Urban Development Corporations (UDCs) derived to large extent from their clarity of purpose. They were focused specifically on physical and economic renewal (though again this was in many ways a limitation) and the outcomes they achieved were perhaps in large part due to this, as well as to the power conferred upon them by their organizational structure, since they were free-standing organizations with planning powers. Hence the Urban Regeneration Companies in both England and Scotland (as described in Chapters 3 and 4) are attempting to apply a similar clarity of aims, clear commitment from key decision-makers, and a tight geographical focus, albeit with a more partnership-oriented approach than that applied by the original UDCs. It may also be argued that the private sector governance model of the UDCs and URGs allows for greater credibility from the perspective of private sector investors than that associated with public sector-led approaches.

However, the document goes on to propose that clarity and focus of purpose must also be applied in public sector regeneration approaches, as illustrated for instance by the City Challenge initiative described in Chapter 3, which encouraged and applied strong leadership. Equally, however, experience in both England and Scotland has shown the benefits of a strategic approach to regeneration, particularly for the most disadvantaged areas. Community involvement is a key element in such areas, and, again, it is hoped that the Regeneration Outcome Agreements mechanism can provide a means of achieving such a strategic approach (Scottish Executive 2006a).

Partnership working

Notwithstanding the importance of a clear focus and strong leadership, the document asserts the continued need for effective partnership working involving all the relevant stakeholders as well as local communities (Scottish Executive 2006). This is a key theme of this book, and an important lesson from the regeneration experience of Dundee, and it will be revisited in the concluding chapter.

England and Scotland in the 1990s: urban regeneration policy compared

Over time, urban regeneration in Scotland would seem to have evolved into a distinctive approach that relies on specific elements such as the geographical targeting of aid, principles of partnership and empowerment, identification of thematic priorities, and implementation of initiatives within a strategic city-wide framework. The Social Inclusion Partnerships (SIPs), for example, set out in Chapter 8 in terms of the experience of Dundee, emphasised the designation of specific projects within a city-wide regeneration strategy. The SIPs initiative also addressed specific themes, including those of young people or drugs, within the context of the use of partnership to implement policy programmes at the community level.

The consistent emphasis on partnership and capacity building in Scotland reflects what may broadly be described as a corporatist influence. Consequently, it may be argued that 'although there may have been a rejection of corporatist or consensual politics south of the border, in many respects they have survived in some areas of policy in Scotland' (Brown, McCrone and Paterson 1996, 106). While such corporatist tendencies may be seen as centralist, elitist and exclusionary, such an approach can also apply policy experimentation from the centre, based on a broad consensus, which can lead to innovation (Judge 1993). Furthermore, the approach to regeneration policy and practice in Scotland seems to have involved distinctive characteristics such as a lead role for local authorities, a high degree of consensus between central and local government on urban regeneration policy and implementation, and a frequent emphasis on strategy and local community involvement (Roberts 2004). While many of these characteristics are shared in the English context, it may be argued that a degree of distinctiveness persists.

There have, however, been similarities between England and Scotland in terms of policy aims and direction in the 1990s. For instance, the emphases within the SIPs initiative of social inclusion and policy integration closely paralleled the English policy context. Moreover, the evolution of regeneration aims in Scotland in the 1980s and 1990s has much in common with that in England. For instance, the shift in Scotland to principles of partnership and empowerment within a strategic framework paralleled that in England. This is perhaps a result of the broad consensus that emerged in the 1980s regarding the inadequacy of narrow approaches to regeneration, and the perceived need for community involvement, integration and awareness of the strategic context.

Hence the English New Deal for Communities (NDC) initiative was significant in its similarity to the Scottish SIPs initiative, since NDC took forward the government's commitment to addressing problems of social exclusion, in the context of the work of the English Social Exclusion Unit. More specifically, NDC sought to tackle multiple deprivation in the poorest areas by addressing worklessness, improving health, reducing crime and raising educational achievement. Moreover, both the SIPs and New Deal for Communities initiatives involved the selection of target areas that were to produce delivery plans, formulated and implemented by local partnerships. And in both cases, these delivery plans were intended to emphasize longer-term strategy and integration with other government regeneration programmes. In addition, the SEU was paralleled in many respects by the Scottish Social Inclusion Network.

There were other points of similarity in terms of policy application between Scotland and England in the 1990s. In particular, the use of competitive bidding as a key mechanism became enshrined in all aspects of regeneration policy after the 'Challenge Fund' model was set up in the 1990s (Oatley 1995). In addition, there was a strong emphasis on greater targeting of aid to smaller areas in both contexts, which may of course be partly explained by the perceived need to reduce public sector spending in the UK as a whole.

However, some points of divergence between England and Scotland became evident in the late 1990s. For instance, there was an increased recognition in Scotland at this time that those suffering from exclusion in Scotland perhaps tended to be even more concentrated in particular geographical localities and communities than in many other contexts in the UK. While this may be largely in the realm of rhetoric, other differences in the nature and application of policy that emerged after 1998 seem to be of greater substance. For instance, the SIPs umbrella covered a variety of measures and initiatives that were more fragmented in England. Hence, while the SIPs and New Deal for Communities initiatives appear similar in terms of aims, the latter operates in the context of 33 other regeneration initiatives. This led to criticism that in England there were too many such initiatives, with inadequate organizational infrastructure to bring about effective collaboration (Clarence and Painter 1998). The proliferation of area-based initiatives in England seems to have led to particular inefficiencies and confusion at the local level, which have not been paralleled in Scotland. In addition, such initiatives appeared to lead to the distortion of locally-set regeneration priorities, since local authorities and other partners in regeneration were forced to bend their own strategies in order to accommodate the newly-funded elements arising from designation of such zones (Urban Task Force 1999). Again, such problems appeared to be rather less severe in Scotland.

In some respects, it may therefore be suggested that the Scottish approach to urban regeneration has been more effective than that adopted in England. For instance, the Scottish approach appears to have achieved more effective partnership between the agencies responsible for policy and those responsible for implementation (Lawless 1989). Of course, this was facilitated in Scotland by the more manageable scale of regeneration in terms of the presence of a single urban conurbation, and the consequent ability of the Scottish Office to give a greater degree of attention to local affairs than could be afforded by the Department of the Environment in England (Wannop and Leclerc 1987). In addition, the administrative devolution of urban policy to the Scottish Office (prior to the establishment of the Scottish Parliament and Executive) enabled it to act as an effective regional arm of central government (Keating and Boyle 1986). The Scottish Office was also able to work closely in partnership with Scottish Enterprise (the successor to the SDA) in order to provide a co-ordinated one-stop shop facility. Such co-ordination was more difficult to achieve in the English context, as indicated previously.

In terms of the application of targeting of urban policy initiatives to small areas, there were also significant differences in policy orientation between Scotland and England. For instance, in Scotland, there was a long tradition of area intervention, as illustrated by the SDA's 'area projects' referred to above, which was very different to the approach adopted in England. These 'area projects', developed between 1981 and 1994, were geared to economic development, and they involved a concentration of resources into small areas as well as a locally co-ordinated approach that brought together policy and implementation agencies (Boyle 1988; Lawless 1989). Such co-ordination was encouraged by the drawing up of 'project agreements' between the SDA, the local authorities, other public agencies and representatives of local companies (Boyle 1988; 187). While the SDA co-ordinated the actions of the partners, using finance as an inducement (Donnison 1987), it did not take over their role, nor did it prejudice the public accountability of the 'area projects' via the local authorities. Consequently, the 'area projects' largely avoided the 'provoked antagonism' that proved to be such a feature of English urban regeneration mechanisms such as Urban Development Corporations (McCrone 1991).

However, in spite of such 'project agreements', the separation of the planning and implementation functions in the 'area projects' proved to be problematic in some cases (Wannop and Leclerc 1987). Moreover, the use of 'project agreements' was seen to have encouraged a rigid approach, and the introduction of 'self-help' projects marked a shift away from their use (Keating and Boyle 1986). In order to encourage private sector involvement, more flexibility was seen as necessary; this became particularly apparent in the private sector-led approach of later initiatives such as those in Glasgow city centre, which would seem to have marked a convergence with policy in England (Boyle 1988).

More recently, the adoption of social justice as a central principle by government in Scotland, albeit in the context of the acknowledgement of the overriding need to encourage economic growth, would also seem to signify an approach distinct from that in England. This is reflected for instance in the application of mechanisms such as the Community Regeneration Fund. Moreover, while it may be argued that Scottish regeneration policy has lacked clear direction since the creation of the Scottish Parliament in 1999, in comparison with England where there have been several national strategy statements (Tiesdell *et al* 2006), the 2006 *People and Place: Regeneration Policy Statement*, as indicated above, indicates a number of key strategic messages and themes in relation to the national role in regeneration.

Conclusions

The trajectory of urban regeneration policy in Scotland shares many aspects in common with that in England, for instance in terms of the increasing importance ascribed to social inclusion and policy integration in the 1990s, as well as the application of notions of partnership and community empowerment. Nevertheless, there have been important divergencies in terms of urban regeneration policy between England and Scotland, closely linked to the differing administrative, cultural, social, geographical, and economic contexts of these nations. In recent years, a political difference in emphasis has also become apparent, with a greater emphasis in rhetoric (if not always in practice) in Scotland on equity, social justice, and meeting direct and demonstrable social need, for instance in the targeting of funding. Nevertheless such divergences are in the context of an arguably shared overall policy approach which is (in line with parallel debates in the arenas of planning and development) increasingly preoccupied with the prioritization of economic aims (either explicitly or implicitly) over social or environmental aims, with implications for the overall nature, development and application of urban regeneration policy.

Nevertheless, the Scottish Executive's 2006 statement asserts that priorities must be framed with regard to both opportunity and need. However, the extent to which this can be achieved may be questioned, since it is clearly problematic to simultaneously apply criteria in relation to, on the one hand, opportunity-related aspects such as emerging economic opportunities, and on the other hand, need-related aspects such as concentrations of vacant and derelict land. The question may be asked: what is the way forward if (as seems likely) such areas do not coincide? Again, this aspiration to reconcile the potentially-irreconcilable may be likened to the parallel attempt to balance economic and social and environmental policy aims at a broader level, in both England and Scotland. While all are assumed to be important in the context of many policy contexts, particularly regeneration, the promotion of economic growth would seem to be dominant.

Chapter 5

The City and its Region

Introduction

The city of Dundee is located on the east coast of Scotland, on the north bank of the estuary of the River Tay, with the Tay road and rail bridges providing links to the south. In fact, its location within the wider region has proved to be a critical factor in determining its history and evolution in economic terms. This chapter outlines the economic context to the contemporary city, and it locates these changes within a regional context, with reference to the influential Tay Valley Plan, which set the context not only for the city's physical development but also for its social and economic regeneration. This chapter also considers the policy context for inward investment and its effects, and the more recent attempts to plan for the city at the regional and local levels. Finally, it sets out the parallel strategies that exist at the city level, and the changing local governance context.

Figure 5.1 Location of the city of Dundee in Scotland

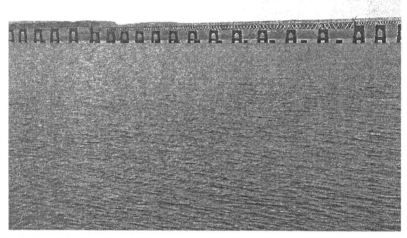

Illustration 5.1 The Tay Rail Bridge

Dundee's early economic history

The location of Dundee ensured that it was well placed to trade within Europe, and it actively traded in hides and wool products, and wine and grain imports. In its early days Dundee was very much a textile town, with a particular strength in the production of coarse wool cloth, plaid and knitwear, which led to the development of processes such as textile dyeing. It also operated as an entrepot port since it received timber and coal products which were transported onwards to smaller ports on the east coast. Indeed, Dundee was Scotland's second highest-value burgh by 1535, though factors such as the Civil War later brought a gradual shift to the west coast ports (Whatley 1991).

In the nineteenth and early twentieth centuries the city's port facilities contributed significantly to its expansion. At this time Dundee became established as an important industrial and trading centre within Scotland. Essentially, the strength of these functions was based upon the manufacturing of goods including shipbuilding and engineering, and ancillary port activities (Whatley 1991). However, the availability of local agricultural produce was also a factor, since the hinterland of Dundee included high quality agricultural land. The economic strengths of the city at this time may be illustrated by the phrase 'jam, jute and journalism': the reference to jam reflects the nearby centres of soft fruits production; that to jute indicates that Dundee was the dominant world centre for the production of this commodity; and that to journalism is based on the importance of the D.C. Thomson publishing empire – the publisher of the *Dandy* and *Beano* as well as the *Dundee Courier*.

Perhaps the jute industry is of greatest historical importance (Whatley 1991). The development of this industry was enabled by the city's experience with textile production. The jute industry contributed substantially to the successful economic development of the city during the Victorian period, and Dundee obtained city status in 1889. Much of the city's tenement housing dates from this period. However, the dominance of this industry itself became a problem, since the city was particularly vulnerable to decline in this sector. In fact, Dundee began to experience economic decline in the 1920s as a result of the global recession, the falling demand for jute-based products, the obsolescence of the industrial base, the high costs of production, and a lack of new inward investment.

Illustration 5.2 Tenement housing in Dundee

The decline in the manufacture of jute is critical; this began to be felt particularly in the inter-war years, and unemployment levels within the city rose to over 70% in 1931-32 (McCarthy and Pollock 1997). This started a process of long-term endemic decline that in some respects continues to the present day, albeit in the context of significant regeneration activity and an expanding service sector. In parallel to these structural economic changes, however, the city was expanding and altering in character since there had been a major expansion of municipal housing in the inter-war period, developed largely in response to national legislation, as well as expansion of the road networks in the city. Both structural and physical issues were addressed by the 1950 Tay Valley Plan, at the regional level, as well as the Dobson Chapman Plan of 1952, at the local level. Both of these are considered below.

The Tay Valley Plan 1950

It is important here to situate the Tay Valley Plan in its historic context. In fact, the 1939-52 period was pivotal in the history of British planning, since the comprehensive idea of physical and land use planning was then embedded in official government policy brought into being by the 1947 Town and Country Planning Act (Ward 1994). The immediate post-war period in Britain saw the creation of a number of radical plans that had longer-term implications for local planning agendas, and the Tay Valley Plan of 1950 set out a very optimistic and ambitious notion of strategic planning for the city region, reflecting many of the ideas associated with regional planning of the period. Like the seminal strategic plans prepared for Greater London in 1944, the Tay Valley Plan articulated the main strategic components of post-war planning orthodoxy, namely decentralization, containment, redevelopment and regional balance.

The significance of context and the force of prevailing ideas cannot be over-estimated. Wannop (1995), for example, identifies the 1940–47 period as the classic vintage with respect to regional planning practice. He suggests that the prevailing economic, social and political circumstances, and the increasing institutional momentum of the time, led to a small number of regional plans being produced that were 'significant beyond anything before and of enduring influence' (Wannop 1995, 7). In his opinion, the regional plans of particular significance were those produced for Greater London and the Clyde Valley. But he points also to a 'last flowering' elsewhere, a process of innovation at a smaller scale of implementation that included the production of the Tay Valley Plan.

Perhaps the most important factor in terms of the origins of the Tay Valley Plan was the legislative change brought about by the 1947 Town and Country Planning Act. The legislation reflected the mood of optimism of the age, in terms of the potential for planning to achieve a variety of outcomes. Thus, from 'the negative and restrictive provisions of the Town and Country Planning (Scotland) Act, 1932, and the Town and County Planning (Interim Development) (Scotland) Act, 1943, the face of the work has changed to *positive creation and inspiring endeavour*' (Lyle and Payne 1950, foreword, emphasis added).

The Tay Valley Plan provides a view of the future of development and growth within the city, in the context of the region as a whole (Doherty 1991). The Tay Valley Region embraced an area that may be termed 'Tayside', which included an area of central Scotland that lies between the Forth and Dee drainage catchment areas. It comprised approximately two million acres, which can be considered in terms of four main physical divisions, namely the Highlands, Strathmore, the Sidlaws and the coastal plain. The Highland area of the Tay Valley formed a dissected plain of marine denudation; the Strathmore area consisted of a wide vale within the syncline north of the Sidlaw Hills; the Sidlaws themselves are formed by the northern limit of a geographical anticline or upfold; and the coastal plain area lies between the River Tay, the North Sea and the Sidlaw Hills to the north (Lyle and Payne 1950). Thus, the Tay Valley not only functioned broadly as a coherent regional unit, but it was also in need of strategic guidelines for development.

The overall aim of the Tay Valley Plan then was to 'form a background for urban and rural reconstruction physically, socially and economically, which must ensue over the next thirty to fifty years in order to roll back the tide of rural decadence and urban concentration' (Lyle and Payne 1950, foreword). The main objectives of the Plan were:

- to ascertain the physical, social and economic advantages and disadvantages of the region;
- to ascertain the physical, social and economic advantages and disadvantages of the region;
- to demonstrate how the advantages of living in the Region may be maintained; and
- to secure the co-operation of the Statutory Planning Authorities in the Region the co-ordination of the various schemes and to assist these authorities to frame their proposals for Development Plans under the 1947 Town and Country Planning Act.

In terms of general principles, the main problem to be addressed by the Plan was the drift in population from the rural areas to the city of Dundee, and from Dundee and the remainder of the city-region to England and abroad. The Plan set out to reverse this drift. It acknowledged that the motivating factor of the drift was primarily economic in character, but also that the issue had strong social and other implications, particularly in terms of the impact on housing. Moreover, the Plan acknowledges the implications of the need for increased government assistance to address these problems.

The Tay Valley Plan was radical in its call for a national solution to such problems, and it suggests that 'a long term national policy is urgently required. Without belittling the benefits bestowed on such places as Dundee by the operation of the Distribution of Industry Act, 1945, let it not be forgotten that this is in effect "stop-gap aid"; it is a short term policy of ensuring full employment and it is far from being a policy which will ensure the optimum distribution of the population' (Lyle and Payne 1950, 235). In this context, the plan builds on, and echoes, the finding of the Report of the Royal Commission on the Geographical Distribution of the Industrial Population (the Barlow Report) and the Report of the Committee of Land Utilization in Rural Areas (the Scott Report). Interestingly, the Plan acknowledges that there 'is no official guidance as to how or when the Government policy will be directed to securing the objectives of the Barlow and Scott Reports, but a gleam of hope can be seen in the Town and Country Planning (Scotland) Act 1947. In this Act ... there are such terms as "the re-location of population and industry" and "the Board (of Trade) shall have regard to the distribution of industry in the National interest". Those outside St. Andrew's House and Whitehall are left to conjecture what lies hidden beneath such phrases' (Lyle and Payne 1950, 243).

The Tay Valley Plan highlights several aspects of the area that were crucial to the longer-term and the sustainable planning for land use in the city-region. For instance, it stresses landscape aspects, together with population characteristics and tendencies, industrial and housing factors, agriculture, accessibility, public utility services, and social conditions. The latter is of note since it explicitly reflects the approach of Patrick Geddes, the influential founder of an important strand of planning philosophy, who retains considerable currency to the present day. The Tay Valley Plan argues that 'Sir Patrick Geddes has defined the objective of social planning as the correct balance and integration of WORK, FOLK and PLACE' (Lyle and Payne 1950, 154), interestingly foreshadowing the contemporary emphasis on 'holistic' regeneration approaches combining economic, social and physical elements. This was an important thread in the strategic framework enshrined in the Plan, which pointed to the social context of the day. Specifically, it indicates that there 'is some dislike of agricultural work, and an unjustified feeling that it is only suitable for the less intelligent youth, combined with a preference for jobs offering cleaner and better working conditions. The great changes which should be brought about in agriculture by the application of more modern scientific methods and extensive mechanisation do not as yet seem to have had any great effect' (Lyle and Payne 1950, 155).

In terms of population, the Plan explicitly argues for a reduction in the size of Dundee, since 'Dundee has reached its maximum desirable population, and in the national interests it is essential to encourage the dispersal of some of the population and industry to the smaller burghs and villages' (Lyle and Payne 1950, 277). Nevertheless, some observers argued for the concentration of a sizeable proportion of the projected increase in the population of the region (Pocock 1968).

In the context of Dundee itself, the Tay Valley Plan contains proposals for the creation of 'neighbourhood units'. It asserts that the concentration of population in towns and cites has led to 'breakdown of neighbourhood feeling' (Lyle and Payne 1950, 241) which has in turn contributed to problems such as traffic danger. Consequently, it was felt that there was need to 'work out some organisation of [a town's] physical form which will aid in every way the full development of community life and enable proper measure of social amenities to be provided and arranged to advantage in each residential neighbourhood' (Lyle and Payne 1950, 241). This led to the concept of neighbourhood units as the rational unit of organization of residential space.

In terms of industry, the Plan aims to maintain a balanced industrial structure for the city. Specifically, it argues that:

> the introduction of suitable light industries would not harm the attractions of the area from the tourist's point of view, would give a wider choice of career to the inhabitants, and in certain cases would help to minimize the season fluctuations of employment experienced in such occupations as agriculture and the tourist trade. Determined efforts are needed to arrest the drift of population away from the country to the towns ... (Lyle and Payne 1950, 57).

Furthermore, the Plan argues that it was necessary to stop the drift of population from the rural hinterland of the city into Dundee, by preventing the introduction of new industry in the city (given that the issue of unemployment appeared to have largely been addressed). In addition, it suggests that there 'appear to be no justifiable or logical arguments why Dundee should grow any larger than its present population of 180,730; indeed, there are many urgent reasons why some 30,000 of the population should be persuaded to move out to make Dundee fifteen neighbourhood units of average 10,000 each, and thus contribute to an optimum distribution of the population' (Lyle and Payne 1950, 278).

As a result of such factors, the Plan broadly recommends the development of smaller and medium sized towns and rural settlements rather than the expansion of the city, in order to ensure a more even distribution of population and a better balance between urban and rural areas. This strategic approach had been envisaged by the Barlow and Scott Reports, which also reflected the radical aspirations of the government of the day. Hence the Plan suggests that 'there is much to commend the expansion of small towns' (Lyle and Payne 1950, 243). Overall, therefore, the Plan proposes the simultaneous development of selected rural centres, the upgrading of selected smaller towns to development centres, and the redevelopment of regional service centres such as Perth and Dundee.

In terms of the means of bring about the Plan's objectives in terms of urban concentration, the Tay Valley Plan proposes a clearly defined green belt in which development would not be permitted. Indeed, it highlights the sharp division between town and country which was prevalent in Scotland at the time, but not to the same extent in England, where there was more evidence of a gradation in terms of suburban development. This, it suggests, was a source of celebration for Scotland, and it goes on to recommend the maintenance of a relatively high density in towns by means of the tenement type of development form.

The Dobson Chapman Plan 1952

The 1947 Town and Country Planning Act enabled local planning authorities to control land use and development within their areas, but, in spite of this Act, many authorities were slow to make use of this enabling legislation by producing local plans; instead they continued to concentrate on issues of development control within their areas. Dundee in this respect was typical. However, the Scottish Office encouraged the local authority to organize the preparation of such a plan, and as a consequence the Corporation contacted Dobson Chapman, a planning consultancy based in Macclesfield, to produce a survey and an advisory development plan for the city. This was produced in 1952. In line with the 1947 Act, the Dobson Chapman Plan accepted the need for, and the possibility of, using a 'master plan' for land use and the zoning of functions. As the Plan suggests:

> The primary objective of planning today is to produce an environment in which all the component parts of the area are related to each other by the regulation of land uses, and the siting and density of buildings so controlled, that all function together as a harmonious whole, and render to each and every section of the community the maximum health, amenity and convenience. To achieve this end, some form of control over development is essential, but in no way does this imply the hampering of legitimate growth. By such control, the evils of haphazard, selfish, sporadic or other ill-considered development can be effectively checked and the possibility of much wasteful expenditure of public funds on the provision of services thereby eliminated (Dobson Chapman 1952, 1).

There were also several social and economic factors that affected the environment for planning in Dundee after the Second World War. In terms of population, between 1931 and 1951 the inner city wards lost population while the outer wards gained population. Moreover, by the late 1930s, the districts of mid-Craigie and Beechwood had been developed as slum clearance estates designed for residents from the central part of the city. In addition, similar districts had been planned and partly developed in such areas as Kirkton and Camperdown.

In terms of economic factors, following the Distribution of Industry Act 1945, Dundee was designated as a Development Area, and industrial estates were subsequently developed on several greenfield sites on the periphery of the city. By the time of publication of the Dobson Chapman plan, a number of new firms had located in such districts, including Timex UK and National Cash Registers. Such firms had been attracted to the city by the grants associated with Development Area status as well as the easy availability of labour (Doherty 1991).

The Dobson Chapman Plan is based on a number of general principles. First, it seeks to preclude the development of the city beyond suitably defined limits, and therefore proposed to create a belt of open space around the periphery. Second, it seeks to define zones for industrial development where existing industry could be consolidated and where a limited number of new industries could be sited. Third, it proposes that residential areas should be consolidated into a series of neighbourhood units, as well as a re-planning of the central area of the city. Fourth, the issue of communications is assumed to require the need to secure the maximum of convenience for the residents of and visitors to the city.

Significantly, the Plan considers the city of Dundee to be unique in one aspect, since it suggests that:

> An outstanding feature of the City of Dundee is the very strong feeling of corporate consciousness possessed by its inhabitants, a most important contribution to which has been the fact that, although the City is completely self-contained and possesses most forms of communal facility, it is yet of a size which allows the Dundonian to feel he [sic] is indeed a member of a cohesive community and not, as is so often the impression in many of the largest cities of Britain – certainly in Glasgow, Manchester, Birmingham and even, perhaps, Edinburgh – a member of a community so large that it is only manageable when sub-divided into what are virtually completely separate and distinct sections, the inhabitants of one having little or no fellow-feeling or community of interest with those of another (Dobson Chapman 1952, 19).

Overall, the Dobson Chapman Plan reflects the optimism of the day in terms of the potential for the role of the statutory planning system: 'By the organisation of town and country planning and economic planning on a national basis it is now possible, by proper control at the national and regional levels, to secure a national distribution and re-distribution of population in order to bring abut adequate balance in the various regional and local economic and sociological structures' (Dobson Chapman 1952, 25).

Issues of population underpin the principal aims and concerns of the Plan, as in the case of the Tay Valley Plan. However, while the Dobson Chapman Plan of necessity gave less attention to the regional implications of population shifts, its conclusions were in direct conflict with those of the Tay Valley Plan. Specifically, the Dobson Chapman Plan suggests:

> The fact that the total foreseeable population plus a limited further addition, could be catered for within the city boundary is of considerable significance, in that the results of our more detailed survey of the city has indicated that the need for the planned dispersal of population into near or distant areas within the adjacent counties and outwith the city boundary, as envisaged by the Tay Valley Plan report need not in fact arise, nor do we consider that the adoption of the drastic ... process of decanting the population, with or without attendant industries, is in the least desirable (Dobson Chapman 1952, 62).

The Dobson Chapman Plan highlights the need to maintain the city as a self-contained entity, thereby precluding the extension of development beyond defined limits, and it proposes the creation of a green belt around the city as a means of achieving aims for urban containment. Overall, the Plan assumes the need for a city population of 180-185,000. In terms of residential areas, however, the Dobson Chapman proposals reflected those of the Tay Valley Plan in that they involved the creation of neighbourhood units. Such Units were conceived, as in the Tay Valley plan (but not acknowledged as such) as a three-part hierarchy with a balanced social mix occurring at each level. Basic Units were conceived as accommodating around 1,500 people, with several such Units forming a neighbourhood unit with a population of up to 15,000. District units were to be formed by two or more neighbourhood units. In terms of detailed design, so-called 'Radburn' principles of design were applied so as to separate pedestrian from vehicular use.

The Dobson Chapman Plan clearly acknowledges the economic basis for the problems of population loss within the region. This was because the jute industry had been adversely affected by competition from India, which led to a response in terms of the designation of the City as a Development Area under the Distribution of Industry Act 1945. The result was that a large number of new industries were established, mainly adjacent to a new industrial estate on the outskirts of the city. The Plan considered such action to be adequate since 'the City has now had its transfusion of new blood and the process of rejuvenation must be directed elsewhere if the City is not to suffer from blood pressure, hardening of the arteries and other ailments likely to affect its somewhat ageing body' (Dobson Chapman 1952, 37).

In terms of spatial implications, the Dobson Chapman Plan assumes the necessity to designate zones that could consolidate existing businesses, contain new ones, and provide space for the relocation of businesses located on unsuitable sites. This was in line with contemporary planning theory which assumed the need to segregate uses. In fact, the Plan recognizes three types of industrial zones, namely older industrial areas such as Blackness, new industrial zones such as the Kingsway belt, and other areas comprising light industrial uses which were judged to be more compatible with adjacent residential areas.

The Dobson Chapman Plan considers the detailed planning of the city in more detail than the Tay Valley Plan. The implications, for instance for the central area of the City, are interesting to note in the light of more recent developments and the contemporary focus on conservation. For instance, the Plan notes that 'the planner naturally endeavours to retain as much of the existing [land use] pattern as possible.

In particular does he [sic] retain in his scheme as many of the existing buildings and streets as he reasonably can, and maintains a watchful eye over buildings and objects of historical and architectural merit or interest value' (Dobson Chapman 1952, 118).

In spite of this consideration of the need for careful conservation, the Plan gives little emphasis in practice to conservation, and statements of support for this conflict with the overall emphasis on redevelopment. Indeed, the Plan proposes a fundamental redevelopment of the central area of the city to incorporate a new civic, cultural and recreational area to be formed by the reclamation of the older docks. This area was to incorporate a new civic theatre, concert hall and exhibition hall as well as provision for 'lighter forms of entertainment and recreation' (Dobson Chapman 1952, 94) including a boating club and restaurant. In addition, the Plan proposes the redevelopment of the University area by means of a proposed "processional route", comprising an ambitious, tree-lined 'processional way' leading from the University to a proposed new square in the city centre. While such proposals did not in fact materialize, the subsequent redevelopment of much of the city centre for new shopping uses broadly followed the Plan's suggestions.

In terms of the central area, the plan applies the concepts of functional zoning referred to above. Functional separation of uses is seen to be necessary to solve the problems for the central area such as congestion, public transport and location of public transport terminals. The main proposal is for the construction of an inner ring road that could filter away traffic from the central retail and commercial area. The aim in terms of communications is therefore for maximum traffic efficiency. While this may seem a radical proposal, the idea of an inner ring road had in fact been proposed by James Thomson in 1918.

The Dobson Chapman report allowed for public scrutiny in the production of the city development plan, which was submitted to the Secretary of State in 1956, and a revised Plan was approved in 1959. In line with the requirements of the 1947 Act, a five year review of the Plan was conducted in 1964, with a revised Plan issued in 1966. Following this revised Plan, however, doubts began to arise as to the continued relevance of the 1947 Act. A new Town and Country Planning Act was passed in 1969. Nevertheless, the subsequent Development Plan review in 1971 made no new recommendations for the form of development planning in Dundee (Doherty 1991).

The Dobson Chapman Plan had a profound impact on the design and organization of Dundee in subsequent decades. The Plan provided guidance for development control, prior to submission and acceptance of the official city plan, which subsequently embodied many of the recommendations of the Dobson Chapman Plan. Partly as a result, major shifts took place in the city from 1945 to 1975, with the demolition of the 19th century heritage and the dispersal of population and industry away from the centre of the city, to the periphery. Moreover, several specific recommendations of the Plan were taken up, albeit in an amended form. For instance, proposals for the Blackness area were carried forward to an extent in the 1959 City Development Plan, and the area subsequently became a Comprehensive Development Area in 1961. Similarly, the Overgate area was redeveloped as a shopping centre, as suggested in the Dobson Chapman Plan, together with the refurbishment of the Wellgate area as a further shopping centre. These proposals were also incorporated in the City Development Plan of 1956 and its later amendments.

However, many of the Plan's recommendations were undermined by lack of concern for the details of implementation. For instance, it may be argued that the peripheral housing estates which followed the prescriptions of the Plan, were built too quickly, with insufficient attention given to standards of construction. In addition, there was a lack of provision of communal facilities in these estates, as well as insufficient social mixing, with the arguably simplistic assumptions of the Dobson Chapman Plan being unable to be translated into reality. Of course, the effects of these factors were later felt, and continue to be felt, in the problems brought about by the development of peripheral housing areas such as Whitfield, described in Chapter 7.

Furthermore, the development of the ring road – a critical recommendation of the Plan – was substantially delayed, only to be completed some 40 years later. Moreover, there would seem to have been a lack of recognition in the Plan of the need to take into account contextual factors such as the likely impact of the Tay Road Bridge, which was to have a major impact on the city from the 1960s onwards (Doherty 1992). Similarly, the Plan seems to have given insufficient attention to broader contextual factors, since it was based on the assumption of full employment, with the main problem anticipated as the need to contain population increase. Consequently, as Doherty (1991, 30) suggests, 'the Report thereby underestimated the decline of Dundee's industrial base and was unable to anticipate the emergence of a 'post-industrial' city. A plan devised to contain and even inhibit growth was simply unable to cope with manufacturing decline and became increasingly irrelevant as the post war boom waned'.

Subsequent planning and regeneration

Notwithstanding the problems of the 1950s, the period from 1945 to 1970 in Dundee brought considerable diversification of the local economy, partly resulting from the financial incentives and provision of industrial sites that resulted from the terms of the Distribution of Industry Act 1945. Hence a number of northern American companies located in the city during this period, and the light engineering, electrical engineering and associated sectors expanded. However, by the 1980s many aspects of deep-rooted economic decline were again evident, due to the effects of technological change and overseas competition (McCarthy and Pollock 1997).

In physical terms, the Tay Valley and Dobson Chapman Plans provided the basis for extensive peripheral public housing development, and, together with a 1959 Development Plan, they encouraged the adoption of Comprehensive Development Areas to address the issues of substandard housing and industrial dereliction in the inner city. However, it would now appear that the CDA programme fuelled the problems of inner city social dislocation as well as a legacy of partly cleared plots and vacant land in many areas (McDougall 1993). In addition, the piecemeal development of the central area of the city was of poor quality in both functional and aesthetic terms. This, combined with the structural economic problems of the city, served to exacerbate the poor image of the city, which was based then upon overcrowded Victorian tenements, multiple deprivation and widespread physical dereliction. While much has been done to address these issues, the legacy of a negative perception based in part upon the city's history continues to this day, and this is one of the reasons for the resources applied to city marketing campaigns, and new flagship development schemes, intended to 're-image' the city.

Following local government reorganization in 1975, the newly-created Dundee District Council attempted to simultaneously improve the condition of the peripheral housing areas and to address the physical problems of the inner city. These aims were to be achieved via the early preparation and adoption of a local plan for the inner city (Dundee District Council 1979). This area covered the whole of the pre-20th century area that was contained between the outer ring road (the 'Kingsway') and the waterfront, but excluding the city centre. The Town and Country Planning (Scotland) Act 1972 required the preparation of local plans, and the Inner City Local Plan was also seen as a catalyst for urban regeneration. The plan's proposals included an extensive programme of environmental improvements, supported by the Scottish Development Agency. These, together with housing improvements, demonstrated the commitment of the local authority to the area (Pacione 1985). This process provided a valuable incentive to private sector investment, culminating in many projects such as the conversion of the Upper Dens Works to residential use, and the redevelopment of the Camperdown Jute Works complex at Lochee to create retail and residential uses (McCarthy and Pollock 1997).

Local government reorganization, national planning and city regions

In April 1996, as a consequence of the Local Government (Scotland) Act 1994, the two tier structure of regions and districts was replaced by a unitary system of Councils. The legislation introduced a streamlined, single-tier, market-oriented, enabling system of local governance in which the new Dundee City Council had a range of defined responsibilities including structure planning, local planning and development control, social services and economic development. The principal function of the Council therefore became to enable the delivery of defined services rather than assuming responsibility for delivering these, and Tayside Regional Council disappeared, with a new City Council replacing both it and the previous District Council.

It is relevant to consider here the responsibilities of Scottish Enterprise, which involves the integrated delivery of economic and business development initiatives, the provision of training and the implementation of measures to secure the improvement of the environment in Scotland. In formal terms, Scottish Enterprise is charged with the responsibility of stimulating self sustaining economic development and the growth of enterprise, securing the improvement of the environment, encouraging the creation of viable jobs, reducing unemployment and improving the skills of the Scottish workforce. The delivery of its integrated enterprise and training services is sub-contracted by Scottish Enterprise to a network of Local Enterprise Companies (LECs). LECs are not statutory bodies per se but are private companies constituted under the Companies Act 1985 to bring a direct knowledge and understanding of the needs and opportunities of the local economy to the delivery of the government's programme for enterprise, environment and training, and to engage the private sector directly.

Scottish Enterprise Tayside (SET) is the LEC with responsibility for Dundee and is a partner within the Dundee Partnership (considered in detail in Chapter 8). It provides a delivery framework for training, enterprise and business development and environmental improvement in Tayside. The training programmes comprise national schemes and the design of customised measures to reflect local circumstances throughout Scotland. SET implements government-funded training programmes in their individual areas, assesses local requirements for industrial property, environmental improvement or land renewal schemes, and investigates and recommends new local initiatives for training and economic and social development.

It is also of relevance here to note the aspirations and policies of national government and their effects on Dundee and the city-region. Broadly, while the growth of the national economy would appear to be paramount, there is a parallel emphasis on the need for social justice as a fundamental right (Scottish Executive 2003b). Moreover, it may be argued that the Scottish Executive's emphasis on sustainable development is incompatible with poverty, social alienation and urban dereliction (Lloyd *et al* 2006). Nevertheless, the Cities Review (Scottish Executive 2003c) highlighted the primacy of cities in Scotland as economic engines of growth, as well as the need for enhanced competitiveness, regional balance, social and community justice and environmental sustainability. The importance of cities in Scotland is also reflected in key policy documents of the Executive such as the *Framework for Economic Development* (Scottish Executive 2004a) and the *National Planning Framework for Scotland* (Scottish Executive 2004b).

The latter is particularly significant since it provides a clear strategic spatial planning framework for public policy as a whole, including all aspects of regeneration activity. It also draws particular attention to the implications of the enlargement of the European Union and the European Spatial Development Perspective. It indicates clearly that the role of cities as key drivers of economic change is critical, and it sets out a number of key policy ambitions for Dundee. In particular, it indicates that the main challenge for the city is to reverse population loss, though it also acknowledges that much progress has been made in improving the environmental quality of the city centre, enhancing cultural facilities, and developing clusters of the knowledge economy. While it acknowledges that the city has been successful in attracting increasing numbers of students to study in the city, the problem remains that many of these choose to leave after their period of study. The priorities for the city, the Framework indicates, include the redevelopment of the waterfront area, further development of knowledge economy clusters such as the Digital Media Campus, the Technology Park, the Medipark, and the Scottish Crop Research Unit, including the improvement of public transport to these areas. It also suggests that the attraction of the city as a focus for jobs would be significantly improved by the reduction of the rail journey time to Edinburgh, possibly to under one hour.

In terms of the city region scale, the *National Planning Framework for Scotland* proposes a clear case for the management of the spatial economy of Scotland by means of defined city regions. Hence Dundee can be seen to be linked to its functional housing and labour markets in a way that is difficult to reconcile with its very tightly drawn local authority boundary. Moreover, the proposals for modernization of spatial planning in Scotland, culminating in the Planning (etc.) Scotland Act 2006, involves the replacement of structure plans in Scotland by 'city region plans' required only for the four largest cities (Scottish Executive 2004). Interestingly, this is in many ways an endorsement of the strategic planning approach applied by the Tay Valley Plan.

In terms of more specific visions for the future of Dundee, the Cities Review (Scottish Executive 2003c) reflected the views expressed in the Scottish Executive's *Partnership Agreement* (2003b) that individual cities in Scotland should prepare growth strategies to maximize the potentials of the unique mix of characteristics and opportunities. Hence the Executive suggested that each city should draw up a ten-year 'city vision' that would be used as the basis for attracting funding via a Growth Fund (Scottish Executive 2003c). Significantly, such 'city visions' were to be taken forward via existing partnerships and collaborations between the main stakeholders active in the city economy, such as councils, community planning partnerships, and communities themselves. The resulting city vision for Dundee reflected the difficulties associated with the mismatch between the city's administrative boundaries and its functional boundaries. It also emphasized the importance of partnership and collaborative working in the city, as well as community planning. The vision itself was based essentially on the redevelopment and regeneration of the waterfront area of the city, since this was seen as the most important focus of regeneration in the city.

Conclusions

The recent and contemporary regeneration activity of the city of Dundee can be seen to be the consequence of a long-term process of industrial decline and post-industrial restructuring. However, this has been combined in Dundee with an assertive and innovative approach to the management of change. This is illustrated for instance by the 1950 Tay Valley Plan, the strategic approach of which would seem to be endorsed by the recent focus of the Scottish Executive on planning for the four city-regions of Scotland. The Tay Valley Plan was accompanied by the Dobson Chapman Survey and Plan of 1952, which set out a strategic framework for Dundee. Although there is an important strategic-local relationship between the documents, there is also a fundamental point of divergence between them. Specifically, differences emerged in relation to attitudes to population dispersal, and in the implementation of planning at different levels. In terms of effects on contemporary regeneration issues, the emphasis of the Dobson Chapman Plan on urban containment together with development of peripheral neighbourhoods would seem to have led to the subsequent problems of peripheral housing estates as indicated in Chapter 7, though such problems may be seen to have arisen largely as a result of inadequacies of implementation. In retrospect, both plans failed to reflect the complex reality of external factors, but the Tay Valley Plan in particular has contemporary resonance for strategic planning for city-regions in Scotland.

In terms of the broader context, there remain problems with the institutional context for regeneration in Dundee, not least the local authority boundary, which fails to reflect the more meaningful notions of functional travel-to-work areas or the broader city-region concept. However, the resurgence of interest in city-region approaches as indicated by proposals for city-region plans in Scotland indicates an acknowledgement of the need for consideration of functional urban areas in spatial planning and regeneration policy.

Chapter 6

Property-Led Approaches

Introduction

This Chapter sets out the origins and development of specific property-led approaches to regeneration in Dundee, which are linked to previous initiatives such as the Dobson Chapman Plan, as set out in the previous Chapter. It considers a series of projects that, together, illustrate the problems and potential of the property-led approach to regeneration. These projects illustrate significant sub-themes within the field of property-led regeneration, and they comprise the Blackness Industrial Improvement Area (industry-led regeneration), the Central Waterfront scheme (tourism-led regeneration), and the Dundee Contemporary Arts Centre (culture-led regeneration). The implications arising from effects of each of these are discussed, and finally the implications of the 'property-led' approach to regeneration that links these initiatives are considered.

Industry-led regeneration: the Blackness Industrial Improvement Area

Following local government reorganization in 1975, Dundee District Council embarked on the early preparation of a local plan for the inner city (McCarthy and Pollock 1997). The area covered by the plan included all of the pre-20th century part of the city that lay between the outer ring road (the Kingsway) and the waterfront, but excluding the city centre. The Council saw this plan explicitly in terms of its potential as a catalyst that could bring about objectives for urban regeneration, and it proved to be particularly valuable in indicating the local authority's commitment to the regeneration of this part of the city (McCarthy and Pollock 1997). This was because the plan prompted a programme of environmental improvements that was supported by the Scottish Development Agency and which clearly improved the image of the city to potential investors and developers. By increasing the confidence in the city, this plan contributed to the eventual development of projects such as the award-winning joint venture residential scheme for the conversion of a former jute mill at Upper Dens Works, as well as the private sector-led redevelopment of the Camperdown Jute Works at Lochee (McCarthy and Pollock 1997).

However, perhaps the most significant regeneration project that was promoted by the Inner City Local Plan was the Blackness Industrial Improvement Area. This was designated as Scotland's first Industrial Improvement Area (IIA), under the 1978 Inner Urban Areas Act, and it successfully brought together a range of local agencies with an interest in regeneration. Specifically, in 1979 the designation of the IIA was followed by the setting-up of institutional arrangements involving the Scottish Development Agency, Tayside Regional Council and Dundee District Council. These 'partners' formed a project team that was supported by the local business community (Bazley 1993).

The Blackness area comprised a largely industrial complex of buildings to the west of the city centre. This area had been in decline for many years, as indicated by evidence of physical dereliction, and in 1979 it was characterized by narrow streets, many multi-storey buildings in urgent need of repair, and an overall degraded environment (Bazley 1993). The area was around 45 hectares in size, and it had provided an important concentration of industrial activity in conjunction with Dundee's jute industry. It contained in 1979 over three million square feet of industrial and commercial floorspace, though much of this floorspace was located in stone-built, multi-storey buildings that were unsuitable for modern industrial uses. Moreover, while there had once been a substantial residential population, this had been decanted by 1979 to other areas of the city, particularly the peripheral housing estates.

Blackness IIA/Blackness project

The strategy for the Industrial Improvement Area was essentially to direct attention to key sites, remove evidence of dereliction and thereby generate increased confidence and investment in the area. The mechanisms adopted included grant schemes that were set up to assist the refurbishment of property in the area, and business advice that was provided for industrial enterprises (MacDougall 1993). In addition, alternative uses were sought for redundant buildings, including use as premises for small businesses, and new traffic circulation systems were progressed in order to modernize transportation infrastructure, including the development of a new link road as well as improvements to street lighting and pavements. In addition, buildings or sites that formed key gateways into the area were upgraded, and some new industrial units were provided. As a consequence of such activities, the physical decline of the area was largely halted. Furthermore, the existing enterprises in the area were encouraged to stay with the result of the safeguarding of jobs, and some new employment creation.

In terms of the detailed methods adopted, it was again the institutional arrangements that were particularly significant. The co-ordination of the key partners was achieved by means of the Blackness Project Agreement, which provided a basis for consensus and co-ordination, and allowed the three partners to contribute their most valuable skills. Hence the Scottish Development Agency provided valuable business advice, and Dundee District Council provided design expertise for environmental improvements, as well as using its power of compulsory purchase to assemble sites which for a range of uses. Both the District and Regional Councils also promoted a property grant scheme that involved giving advice on property in the IIA area as well as assessing the schemes that were submitted for consideration for grant aid (MacDougall 1993).

In addition, innovative mechanisms were adopted for the physical regeneration of the area. For instance, the important potential role of public art was recognized, with several examples initiated as a result of the IIA. These came about because of the creation of the Blackness Arts Project, which provided a public arts input to number of environmental improvements in the Blackness area (Bazley 1993). In terms of objectives for renovation, the mills provided a particular problem since, while obsolete for many potential uses, they were all listed. However, several were brought back into re-use by the application of a flexible land-use policy, and resulting new uses included the provision of student flats. In addition, clearance and landscaping of the road reservation provided a major environmental improvement, and this proved to be a major factor in the decision of the Nationwide Housing Trust to develop the site at Coupar Street for private and co-ownership housing. This was in fact the first non-public sector housing project in the inner city.

In terms of economic development, the initial aim of significantly increasing the level of employment in the area unfortunately became unrealistic as a result of the closure of the area's main employer, the Caird Carpet Works, shortly after the creation of the Blackness Project. This was compounded by the steady decline in the level of employment of the other large companies in the area during much of the life of the Project (Bazely 1993). However, there were other successes. For instance, the Dundee Industrial Association was set up as a local property developer and industrial enterprise agency. The Blackness Trading Company was also established. It was supported by the IIA project partners as well as the Manpower Service Commission, a local trust and a series of private companies, and it offered work experience to school leavers in its own workshop as well as in other private firms in the area (MacDougall 1993). As a consequence of such initiatives, the Blackness Project stabilized the level of employment in the area and significantly increased the number and range of enterprises operating.

The Blackness Project became part of the Dundee Project in 1982, though initiatives in Blackness continued to be co-ordinated from a local office in Blackness until completion of the Blackness IIA Project in 1984. The property grant scheme, however, continued to operate until 1986.

Evaluation

In physical terms, the Blackness Project clearly provided a range of environmental improvements as well as the basic infrastructure necessary for development. In economic terms, a review of the project confirmed that it had been successful in halting the economic decline that had been experienced in the area, though the overall net increase in employment was modest in the context of the decline in manufacturing industry in the city as a whole. The Blackness Project received a commendation from the RTPI (Scotland) and the Saltire Society in 1986. Consequently, the Dundee Project, with its city-wide remit, was set up in 1982 to replicate the kinds of achievements brought about by the Blackness Project, but on a city-wide basis. One of the first sites prioritized by the Dundee Project was the Central Waterfront site, considered below.

Tourism-led regeneration: Dundee's Central Waterfront

The port function in Dundee developed over a long period, though it declined significantly in the 20th century. Dock development began around 1815 (Corporation of Dundee 1957), and the last dock built at Dundee was the Fish Dock, opened in 1900 (Corporation of Dundee 1957). Significant decline in the jute industry followed, which had major effects on the city's economy (Whatley 1991; Doherty 1991). In fact, the ideas of James Thomson, the advocate of extensive planning intervention who became City Engineer and Architect in 1906, led directly to proposals for development of redundant areas of the waterfront (Braithwaite 1989). Specifically, Thomson proposed the construction of a Civic Centre on part of the waterfront as the basis for the long-term improvement of the whole waterfront area (Subedi 1993). This idea was put forward in the Town Planning Report submitted to the Town Council in 1918, which recommended a phased programme of restoration work to facilitate redevelopment (Subedi 1993). However, Thomson's proposals were never implemented.

The silting of the Tay began to prevent the new generation of large vessels from entering Dundee's harbour after the Second World War (Doherty 1991), which increased the need for re-use of much of the waterfront. Furthermore, plans for a bridge further east were accepted by the city in 1960. The most important factor in this siting decision was cost, since the shortest route, downstream of the port, was more expensive; in addition, the route chosen had the cost advantages of using the tidal harbour on the Dundee side, which was clear of expensive underwater cables. However, the bridge bisected the waterfront area and therefore reduced the potential for the consideration of the area as an integrated whole, and this has continued to act as an obstacle to effective regeneration of the waterfront area (McCarthy 1995).

Dundee's Enterprise Zone

Partly as a result of the evidence relating to the work of the Dundee Project's partnership approach, the city was successful in its bid for an Enterprise Zone (EZ) for Dundee (Bazley 1993). Enterprise Zone designation was seen as a means of increasing the city's chances of attracting inward investment (Henderson 1994), and a bid for enterprise zone status was made in 1982, with the Tayside Enterprise Zone declared in February 1984. It consisted of six sites, including a former shipyard site in the port area as well as the Technology Park and Central Waterfront areas. Each site was different and was therefore targeted for different uses. For instance, the port site was considered suitable for development of the offshore oil sector, and specialized oil rig conversion and related work was attracted after investment on the site by the Dundee Project partners as well as others (Bazley 1993). By contrast, the Technology Park was considered to be suitable for a high-quality mix of high technology industrial uses. The Central Waterfront EZ site was seen as appropriate for commercial, tourism and retail functions, largely because of its central location.

The Central Waterfront EZ site comprised around 12 hectares of former rail sidings and unused ground. This area was close to the city centre, and located between the Tay Road and Rail Bridges (Bazley 1993). However, it formed part of a larger area of around 22 hectares of underused waterfront land. The City Council assumed that a substantial external allocation of funding would be required in order to attract private sector development interest to any part of this larger area (Taylor 1990).

In view of the strategic significance of the site, which was then considered as potentially the most attractive development site in the city, the SDA, via the Dundee Project, commissioned a feasibility study for development of the area. Its report recommended development of the area for a mixed leisure/commercial/residential scheme together with a tourism 'flagship' as a focus (Taylor 1990). This was similar to other waterfront development schemes in many other contexts (Falk 1992; Tunbridge 1988; Law 1988; Brownill 1990). In the Central Waterfront case, the aim of the Dundee Project partners was that public sector partners would assemble the land and install infrastructure, while private sector partners would provide finance and project management expertise (McCarthy 1995).

Within the Central Waterfront EZ site as a whole, the Stakis Earl Grey Hotel was the first scheme to be completed. This site was owned by the City Council, but nevertheless, few design constraints were imposed, and the scheme proved to be generally unpopular (a view which remains to the present day). This prompted public fears about the remainder of the Central Waterfront site (McCarthy 1998a).

It is significant in this context that the SDA's 1984 Development Brief for the Central Waterfront site contained a number of potential contradictions and inconsistencies. In particular, it indicated that certain types of retailing would be considered favourably, since the intention was to complement the leisure-oriented approach of the Brief's proposals (McCarthy 1995). However, food superstores were not explicitly ruled out, in spite of the clear indication in the Brief that such uses were undesirable in planning terms. Partly as a consequence, there was the subsequent development of a Tesco superstore on a major part of the site. After publication of the Brief, a partnership was formed between the Dundee Project, GA Properties Limited (developers) and the National Leasing and Finance Company (financiers), who formed a single-purpose joint venture company, Discovery Quay Development Ltd., which was intended to bring about the development of a site on the central waterfront which it called 'Discovery Quay', comprising almost all the EZ site.

The company submitted an application for planning permission in 1986, for a scheme that incorporated a large retail element together with a leisure proposal, but no residential component. It was necessary to submit a planning application because the Waterfront EZ site was subject to a planning scheme that reserved certain developments from the permitted development status of the Enterprise Zone (Henderson 1994). In particular, retail facilities with a gross floor area of over 400 square metres were 'reserved' in this way, and so were not allowed as 'permitted development' (McCarthy 1995).

Nevertheless, a particular area of concern of the planning authorities was the potential for harmful competition arising from the retail proposals, and there were also fears that the leisure uses might only be viable in the short term. Hence Tayside Regional Council 'called in' the application (to allow it to make the final decision).

Nevertheless, the District Council indicated their acceptance in principle of the proposal, and conditional approval was granted in June 1987.

Unfortunately, several of the partners subsequently withdrew from the scheme, following a downturn in the market for leisure uses, partly because of changes in the ownership of some of the leisure interests. However, the proposal for the 'Discovery Point' visitor centre at Craig Harbour, remained, since it was planned as the centrepiece of the site. Indeed, the 'RRS Discovery', Captain Scott's Antarctic Research Ship, was brought back to Dundee, where it had been built, and it was berthed at Victoria Dock so it could be easily transferred to the central Waterfront site (McCarthy 1995). The financial viability of the visitor centre scheme was ensured by a substantial funding contribution resulting from a development agreement involving the retail uses on the waterfront (Henderson 1994). These retail uses comprised a 64,000 sq ft retail unit with an adjacent petrol filling station, occupied by Tesco, and a 30,000 sq ft non-food retail warehouse, occupied initially by Texas Homecare (Bazley 1993).

Illustration 6.1 The RRS 'Discovery', berthed in Dundee's Central Waterfront

Consequently, it was necessary to review the proposals for the Central Waterfront in 1990, though this was made more difficult because contextual factors were in a state of flux. Specifically, the SDA was replaced by Scottish Enterprise National (SEN) in 1990, at the same time as the Training and Enterprise Councils were created in England. Scottish Enterprise National was formed by the merging of the functions of the SDA and the Training Agency. Its main objective was to encourage private sector-led investment and training (Lloyd 1992), largely by means of a network of Local Enterprise Companies that would engage in a variety of economic development activities, and the local enterprise company for Dundee was Scottish Enterprise Tayside (SET), created in April 1991. Because of these changes, there was a significant turnover in staffing within the agencies involved in the Central Waterfront scheme, including the project manager for the waterfront scheme (McCarthy 1997).

Illustration 6.2 HM Frigate 'Unicorn', launched in 1824 and berthed in Dundee's Victoria Dock

A further factor which complicated the situation at this time was the indication by the Port Authority in July 1991 that land might be available for development at the Victoria Dock. In fact many observers had considered Victoria Dock to be preferable to Craig Harbour as a site for a 'flagship' visitor attraction, in spite of the separation of the Dock from the Central Waterfront by the road bridge. This was because Victoria Dock had a more sheltered site and incorporated buildings of historic importance that could be adapted and re-used, as in many other successful waterfront regeneration schemes (Falk 1992). It also contained the historic Frigate 'Unicorn'. Consequently, the SET board suspended consideration of the Central Waterfront site for the visitor centre until the alternative possibility of Victoria Dock had been investigated. However, in practice, the development of Victoria Dock was found to be problematic, because of the presence of remaining tenants with unexpired leases, and the need for extensive infrastructure spending. Hence Victoria Dock was not seen as a viable alternative (McCarthy 1998a).

As a result of these events, the project partners produced a new concept for the Central Waterfront, since they accepted that the leisure uses previously put forward were unsuitable. A market study concluded that the hotel market involved good rates of return on investment at this time, and it was also suggested that office uses could be used to attract further inward investors, perhaps including government departments that were considering relocation. A bar/restaurant use was considered desirable since it could complement other uses, particularly since the visitor centre proposal did not include catering facilities. The new preferred mix of uses therefore incorporated all these elements. The implementation of the 'Discovery Point' visitor centre building progressed in spite of difficulties regarding design details, and it opened in July 1993 after the RRS Discovery had been moved to Craig Harbour. After the development of the visitor centre, a 40-bed hotel was built on the riverfront adjacent to the visitor centre, and subsequent development of a mixed use scheme took place at nearby City Quay (McCarthy 1998a). Environmental improvements have also been made to the waterfront area.

Illustration 6.3 Residential development at 'City Quay'

Illustration 6.4 Environmental improvements in Dundee's Central Waterfront

The Central Waterfront Vision and Masterplan

The 2005 Community Plan for Dundee indicates that the £270 million project for the waterfront of the city, which is to take place over a 15 year period, involves the extension of the city centre to the waterfront area; the creation of a new street pattern based on a grid; improved facilities for sustainable transport; creation of new tree-lined boulevards; provision of new mixed-used developments, formation of a new civic space and dock marina; and a new railway station combined with an new square (Dundee Partnership 2005). The overall scheme is expected to include hotels, leisure uses, office development, and around 400 new apartments, and it is hoped that the new investment will attract over £250 million of private sector investment. The achievement of the project is anticipated to depend upon extensive collaboration between the Scottish Executive, other public bodies and the private sector.

The proposals reflect the Dundee Central Waterfront Development Masterplan 2001-2031 (Dundee City Council 2002). This masterplan was based on a process that started in 1998, when the Dundee Partnership began to consider the potential for re-integrating the Central Waterfront with the city centre. The overall aim was to create a shared vision for the area that would afford a distinctive identity and sense of place, whilst also providing a robust framework for investment over a thirty year period. A Consultants' Report was subsequently produced in 2000 by EDAW, which identified a number of visionary development options. These options then formed the basis for a wide consultation exercise involving a public exhibition and symposium in the Dundee Contemporary Arts Centre as well as other consultative mechanisms such as a joint seminar with the Civic Trust and local branches of the relevant professional institutions. A consensus view gradually emerged on the basis of a masterplan, and, following a further consultation exercise with key agents, a final Masterplan was approved. This Masterplan is now in the process of being implemented, and it was incorporated into a review of the Dundee Local Plan (Dundee City Council 2003b). The kick-starting of the project was made possible by the Cities Growth Fund, and the project is projected to cost around £57 million in total (Scottish Executive 2006), with the recycling of receipts from the sale of sites intended to allow continued funding for the project.

A formal partnership has now been established between Dundee City Council and Scottish Enterprise Tayside to sure the delivery of the project, and a Partnership Board has been set up which incorporates elected member and officer representation from both organizations to oversee the project. An officers' steering group has also been formed to consider the detailed development of the project. While the Tay Road Bridge Joint Board was unable to participate as a full partner due to its legal status, it agreed in 2004 to sell the land in its ownership that was needed for the project to progress, and to remain involved via the Central Waterfront Steering Group (Dundee City Council 2005a).

Evaluation

Notwithstanding the current vision for the central waterfront, any evaluation of the current status of the area must be derived from a consideration of how the original proposals for the area in the 1980s were progressed. In terms of the extent of development initially achieved, approximately 50% of the Waterfront EZ area had been developed by 1994 (Henderson 1994). This may be compared with the Technology Park EZ site in Dundee, which was 100% developed by this time, in part the result of the lower development costs on this greenfield site. In addition, in terms of the objectives sought, the SDA's Development Brief for the area intended it for a mixed leisure/commercial/residential development, focused on a new tourist attraction. It also stated that any retail uses should encourage or complement a leisure-oriented approach to development (McCarthy 1995), because it of the need to avoid the duplication of retail uses in the city centre.

In fact, many of the objectives of the Brief were not achieved. Specifically, there was no residential element, and much of the EZ Central Waterfront site was in fact developed for retail uses initially comprising a retail warehouse (which subsequently became vacant in 2004) and a food superstore, neither of which match the SDA's original aspirations. Furthermore, the SDA's Development Brief stated that it wished to see the Central Waterfront area developed in a comprehensive way rather than being disposed of in piecemeal fashion, to ensure that the overall scheme was coherent. In practice, however, the individual development plots were disposed of in a separate and piecemeal manner, and so the Central Waterfront area became relatively fragmented (McCarthy 1995).

In the initial development of the Central Waterfront, however, designation of the Central Waterfront EZ in Dundee was a crucial factor in the development of the area. However, while EZ designation enabled subsidization of development that might not have otherwise occurred, it also made land-use planning objectives more problematic.

As a result, many objectives for the land-use mix were not achieved (McCarthy 1995). It may also be of significance that residential and leisure developments of the type anticipated for the Central Waterfront were granted permission in other areas of the city. This problem was compounded by the locally-depressed property market shortly after designation of the EZ area. Partly as a consequence, the leisure uses were removed and no residential uses were developed in the area, in spite of the SDA's intentions in the Brief.

Nevertheless, such factors do not seem to wholly explain the nature of development in the Central Waterfront area, because in the Technology Park EZ, development objectives were largely achieved. While the difference in development costs between the two EZ sites was clearly a factor, the priority given to high quality development in the Technology Park site would also seem to have been significant. Here, the City Council ensured appropriate uses and a high standard of design by means of a Section 50 agreement. This specified a requirement for specific types of high technology uses, and indicated that all development would need to be designed to high aesthetic standards within a landscaped, park-like setting. In addition, the SDA installed basic infrastructure and landscaping on this site. This suggests that the contextual economic problems of Dundee did not wholly explain the inadequacies of physical regeneration of the Central waterfront, and that the nature of this area was to some extent the consequence of a series of unconnected, incremental decisions, in terms of both investment and land-use policy (McCarthy 1995).

However, the current Masterplan is clearly attempting to address this issue with a long-term perspective and a new vision for the area that also reflects some of the aspirations of the SDA for the area. In addition, it is interesting to note that many of the elements of the Masterplan, including for instance tree-lined boulevards, and a new civic space, bear close similarity to the development proposals for the area of James Thomson in the early part of the last century. Such proposals, while arguably far-sighted and visionary, were unable to be implemented because of a series of practical problems at the time. This provides evidence, as in the case of the historical legacy of the Tay Valley Plan, of the opportunities to learn lessons from the history of previous proposals, policies and practices.

The final development scheme considered in this chapter is that of the Dundee Contemporary Arts centre, since this illustrates a further 'property-led' regeneration theme – the use of arts and culture.

Culture-led regeneration: Dundee Contemporary Arts Centre (DCA)

Because of the systemic decline experienced in Dundee, as outlined in earlier chapters, there have been a number of attempts to 're-image' the city as part of wider city marketing exercises. For instance, in 1986 a marketing strategy was initiated which was influenced by the earlier campaign of the city of Glasgow, following its 'European City of Culture' designation in 1990. Attempts to 're-image' Dundee have promoted the city's advantages, including high-quality health and education sectors and proximity to some of the most remote countryside in Europe (Lloyd and McCarthy 1999). In addition, such marketing initiatives often emphasized the presence of a core of culture-related facilities including Dundee Repertory Theatre and the Duncan of Jordanstone College of Art, the largest art college in Scotland (McCarthy 2005). Such facilities, together with the presence of a high percentage of socio-economic groups A, B and C1 within a 30 mile radius of the city (Dundee City Council 1996), also contributed to the development of the Dundee Contemporary Arts (DCA) Centre.

The origins of DCA lie, however, in wider strategies to encourage arts and culture in the city (Lloyd and McCarthy 1999). The *Dundee Arts Strategy*, launched in late 1994, aimed to develop Dundee as a regional centre for the arts; to recognize the role of the artist in developing and maintaining the cultural, social and economic fabric of the city; to promote high standards in the creation, presentation and management of the arts; to support and develop the local arts community; and to develop public access to, and involvement in, the arts (Dundee City Council 1996). In 1997 an Arts Lottery grant of £5.38 million was obtained for the creation of a new contemporary arts centre for the city. In addition, a revised arts and culture strategy was set out in 1997, in the form of the *Arts Action Plan for 1998-2000*, which aimed to link arts and cultural initiatives to broader corporate and related policy frameworks for planning and economic development (Dundee City Council 1997). An important element of the plan was its identification of the need for improved physical facilities for the arts, the 'flagship' element of which is clearly the £9 million Dundee Contemporary Arts centre (McCarthy 2005).

The Dundee Contemporary Arts centre was opened by Secretary of State Donald Dewar in March 1999. The function of the building is very broad – essentially to promote the visual arts, crafts and design. It was designed by Richard Murphy, and the building comprises a large gallery space within an old brick warehouse, including two cinemas, a bar and restaurant, and a print studio. In addition, the building houses Dundee University's Visual Research Centre, which is intended to play a central role in the development and use of new media technology. This Research Centre is equipped with £600,000 worth of technology, and its function is to allow artists to produce prints, books and CD ROMs, to create three dimensional models and to make, edit and finish videos (University of Dundee Press Office 1999). Hence the Dundee Contemporary Arts centre combines production and participation by providing a focal point for exhibitions as well as the creation of new work. It also involves an emphasis on access for all sections of the local community. Furthermore, the Centre provides a focal point for the city's 'cultural quarter', situated between Perth Road and the central waterfront area (McCarthy 2005).

Evaluation

In order to evaluate the Contemporary Arts Centre initiative, it is useful to locate it within the broader strategies for economic and social development of the City Council. For instance, Dundee City Council's *Economic Development Plan* (Dundee City Council 1996) outlines the need for promotion of the arts and culture sectors, including entertainment and leisure, in order to boost the city's economy (Lloyd and McCarthy 1999). The Plan aims to promote the improvement of cultural facilities, and it suggests that the cultural industries sector is crucial to economic development in Dundee, since cultural industries have an impact on the growth of other sectors, and they involve high multiplier effects in relation to indirect and induced employment. Consequently, the Plan aims to build on local strengths and opportunities for cultural industries by promoting appropriate business and commercial activities (Dundee City Council 1996).

In addition, the city's Priority Partnership Area/Social Inclusion Partnership strategies involved the four key themes of stability, sustainability, empowerment and prosperity, with detailed proposals targeted to the most disadvantaged neighbourhoods. While these strategies were superseded by the *Community Plan*, for instance, which involves key principles of social inclusion, sustainability and active citizenship (Dundee Partnership 2005), the original aims of the SIPs still remain relevant. Moreover, Dundee's Local Plan (Dundee City Council 2003b) aims to bring about community regeneration by stabilizing communities. In the context of these strategies and aims, it is clear that the Contemporary Arts Centre has provided major benefits. In particular, it has provided an economic boost in terms of the cultural industries, enhanced the city's image to external investors, and provided indirect social benefits for the city as a whole, including areas targeted by the PPA/SIPs initiatives (Lloyd and McCarthy 1999), and now forming part of the city's community regeneration areas.

Nevertheless, there are a number of problems with the DCA scheme as a means of achieving sustainable regeneration. In order to examine these problems it is essential to consider the broader rationale for what has been called 'culture-led regeneration'. This approach has expanded in its application significantly in recent years (Bianchini 1993a; 1993b; Ebert *et al* 1994; Evans and Dawson 1994; Kawashima 1999), largely because of the recognition of cultural industries and cultural provision as driving forces in urban regeneration (Myerscough 1998; Thorsby 1994; Bianchini 1996). Hence for instance the UK Government's Department of Culture, Media and Sport promoted a 'culture-led' approach to regeneration in the form of guidance for local authorities in the preparation of multi-dimensional Local Cultural Strategies (Department of Culture, Media and Sport 1999). Moreover, a previous Secretary of State, Chris Smith, suggested that 'by far the best way of getting social regeneration off the ground in a neighbourhood or a town is to start with cultural regeneration' (1998, 134).

Clearly, such approaches can bring about a number of broad-based benefits. First, they can lead to economic diversification and job creation by means of the expansion of local activity in the 'cultural industries', as well as related areas of activity such as tourism (Booth and Boyle 1993; Bianchini 1993a; Williams 1996). In addition, expansion of the cultural industries sector within cities can, as well as generating external income, act to prevent the leakage of income from a locality (Williams 1997). Moreover, culture-led approaches can promote positive aspects of partnership working, which can in turn facilitate the leverage of external funding (Department of Culture, Media and Sport 1999).

Second, such approaches can contribute to 'place marketing' through image enhancement, which in turn can encourage inward investment (Ashworth and Voogt 1990; Council for Cultural Co-operation 1995). Expansion of the cultural sector can provide an effective means of altering the city's image for those involved in investment location decisions, since cultural facilities are closely linked to perceived 'quality of life' (Williams 1997). Hence cities judged as having a relatively negative image – perhaps even 'pariah' status – have often adopted such strategies with particular enthusiasm (Fitzsimons 1995). Third, increased participation in the arts and cultural activities can lead to increased social cohesion and an increase in the quality of life, particularly for city residents and workers who are marginalized (Matarasso 1997). Furthermore, the development of the evening economy can help to achieve more sustainable development of mixed-use areas (Darlow 1996), and a reduction in crime may also result (Comedia 1991).

However, there are criticisms of a 'culture-led' approach to regeneration. For instance, it may be argued that cultural projects are no more significant than other types of economic development, so that it may be argued they do not merit special subsidies or attention (Bennett 1995). In addition, the concentration of attention and investment on high-profile regeneration by means of 'cultural flagships' in small areas can lead to so-called 'urban boosterism', within which physical improvements are primarily 'symbolic' in value (Harvey 1989; Smyth 1996). It may also be argued that such schemes are largely exclusive in the distribution of their benefits since they are often framed in narrow economistic terms (Paddison 1993; Landry *et al* 1996). As a consequence, peripheral areas of the city may be marginalized, particularly in the case of city-centre cultural initiatives (Williams 1997). In addition, the essential rationale for policy to promote the clustering of cultural uses within central areas may also be questioned (McCarthy 2006).

Essentially, therefore, many culture-led regeneration approaches may have a limited potential for bringing about long-term, self-sustaining solutions to entrenched urban problems (Montgomery 1995). These criticisms of 'culture-led' approaches may also be applied to apparent examples of 'best practice', such as the Temple Bar project in Dublin (Fitzsimons 1996; McCarthy 1998b).

In terms of Dundee's approach to culture-led regeneration, including the development of the Dundee Contemporary Arts Centre as well as broader proposals for Dundee' Cultural Quarter, this seems to only partially meet the objectives that observers such as Booth and Boyle (1993), Bianchini (1993a) and Landry *et al* (1996) set out as a rationale for culture-led regeneration, including the creation of employment and an improved quality of life for the city as a whole. While the creation of the Dundee Contemporary Arts centre is a major achievement (Macmillan 1999), it may be argued that it does not match the broader rhetoric of integrated and inclusive regeneration espoused in the city's Arts Action Plan. Indeed, it may be seen to illustrate an over-reliance on image enhancement and 'symbolic' effects at the expense of wider 'holistic' regeneration objectives (McCarthy 2005). For instance, Williams (1997) suggests that 'outward-looking' strategies may increase, rather than reduce, the leakage of income from the city. Moreover, while access to DCA is free of charge, and the scheme creditably incorporates a broad range of community-focused benefits, it may be argued that its location and amenities do not promote use by those who are most socially excluded within the city.

Nevertheless, there are several different approaches to culture-led regeneration, including for instance 'promotional' and 'integrationist' approaches (Williams 1997). The case of Dundee would seem to illustrate a development of the former, linked to city marketing. While social inclusion has been recently emphasized, the Dundee Contemporary Arts centre may be seen to illustrate the dominance to an extent of a 'promotional' approach, building on previous 're-imaging' initiatives such as the Central Waterfront area development (McCarthy 2005). As indicated above, such initiatives are problematic. Nevertheless, the Dundee Contemporary Arts centre is clearly just one element of a broader approach to regeneration, and its effects must be considered in conjunction with other elements. In particular, the developments taking place in 2005 of additional student accommodation close to the Cultural Quarter, and the creation of Seabraes Yard, a creative media district, on an adjacent site, indicate that the regeneration of the Cultural Quarter and its environs is achieving a greater breadth of use and attraction, which has the potential to contribute to longer-term regeneration outcomes.

New developments

It must nevertheless be acknowledged that physical development in Dundee is in 2006 in a state of flux, for instance with a resurgence in the city's housing market, new private development, student accommodation, and the emergence of new developments related to the city's new economic growth sectors. Of particular note in this respect is the emerging digital media sector. In fact, Scotland as a whole has recently expanded as an important location for digital media businesses, and in 1996 there were 4,500 digital media companies in Scotland. Moreover, Tayside has a significant cluster of digital media businesses with over 300 companies with a combined turnover of over £100 million. Dundee in particular has become home to a number of businesses in the computer games sector. This is partly due to the University of Abertay, Dundee's pioneering role in the teaching of computer games technologies and computer arts.

Partly as a consequence the site of a former railway goods yard has been developed by a partnership between Scottish Enterprise Tayside and private and public sector organizations as a new creative media district named Seabraes Yards. This site is 20 acres in size, and around 360,000 square feet of high quality business space will be created in this area by 2016, to meet the needs of start-up companies (by means of incubator facilities), local businesses and inward investors. The first development on the site was a landmark building (vision@seabraes) comprising a creative media centre with 100,000 square feet of business space As well as computer games and electronic entertainment, allied activities are also planned to be housed in the area. Such activities include film, TV and radio; animation; graphic design, marketing and advertising; music, publishing; software and communications technologies; arts and cultural industries; architectural design and new media. The site will also contain student accommodation and high quality residential accommodation. Significantly, the site of Seabraes Yards, owned by Scottish Enterprise, is adjacent to Dundee's Cultural Quarter, which contains the Dundee Contemporary Arts Centre as well as the Repertory Theatre. The creation of a pedestrian link between these two areas, by means of a series of terraces which drop 18 metres, which is part of the plan for Seabraes yards, will further develop the links between these areas (Scottish Enterprise Tayside 2006).

Conclusions

Approaches to property-led regeneration may be aligned along several different themes. As shown above, examples of relevant initiatives in Dundee have been specifically geared to aspects such as industrial development, tourism and culture. In terms of the timing of such schemes, these approaches transcend any simplistic notion of policy phases, with for instance the Blackness Project starting in the 1970s, and the property-led approach in the Central Waterfront case is set to run for several more decades. However, it is useful here to summarize the essential rationale for property-led regeneration together with its inherent limitations. While recent interpretations of urban regeneration are bound up with notions of 'holistic' regeneration, combining economic, social and environmental effects, approaches in the 1980s in particular were bound up with the concept of regeneration as largely synonymous with property development. This arose from the assumption that such development could bring about a series of catalytic effects. For instance, it was argued that this could bring about the development of environmental advantages by encouraging the development of adjacent or nearby sites; bring about economic benefits by means of job creation; and bring indirect economic benefits as a result of retail or tourism developments.

However, the basic notion of property-led regeneration attracted considerable controversy (Turok 1992). In particular, many commentators questioned the reality of the 'trickle-down' effect noted above, since in practice the benefits of property development seem to accrue to a very narrow section of the population – often not including the local community. Instead, the local community may suffer from the negative effects of development, while reaping none of its benefits. The potentially regressive effects of property-led regeneration are compounded by the nature of development that is often involved. This is because such schemes have often been geared to service sector uses, with beneficiaries including for instance tourists, those with access to cultural facilities, and those able to access well-paid employment in the service sector. Clearly, many local residents may not fall into these categories. Consequently, approaches to 'property-led' regeneration have often resulted in a focus on the high-profile development of small areas with 'islands of renewal' being surrounded by large-scale evidence of deprivation.

Nevertheless, it is clear from the above examples that there are many possible approaches to property-led regeneration, largely depending on the uses under consideration as well as the aims of such approaches. Indeed, partly as a result of the evidence of inequity arising from property-led approaches in the 1980s, urban regeneration is now commonly held to require the combination of explicit social and economic elements in addition to property development. This goes together with the contemporary emphasis in urban regeneration on community involvement and strategic approaches that emphasize sustainable benefits at the city-wide level.

In all the cases of developments outlined above, despite varying aims, benefits would seem to be mainly economic and environmental in character, with social outcomes proving more elusive. Indeed, this has been perhaps the major failing identified in many 'flagship' developments in much of the UK since the 1980s. It therefore seems that property-led approaches, while bringing many advantages, are indeed inherently limited in their capacity to deliver social outcomes. In Dundee, however, more explicitly-social regeneration initiatives have also been applied, for instance the Priority Partnership Areas and Social Inclusion Partnerships initiatives, and subsequent application of the Community Regeneration Fund (as set out in Chapter 8), so these must also be taken into account in an overall assessment of regeneration in Dundee.

The next Chapter considers a larger-scale housing-led regeneration initiative, the Whitfield Partnership, and locates this within the broader historical context of the development of this area.

Chapter 7

Whitfield: A Peripheral Estate

Introduction

This Chapter sets out the experience of the Whitfield area of Dundee. This area represents an important and seminal policy initiative, as one of the four Scottish Housing Partnerships. The Chapter outlines the origin and development of the Whitfield Partnership, and locates this within the broader historical context of the development of the area. The Chapter also assesses the effects of the Whitfield Partnership initiative, and discusses its achievements as well as its shortcomings.

The Scottish Housing Partnerships and *New Life for Urban Scotland*

The 1988 *New Life for Urban Scotland* White Paper (Scottish Office 1988) provided the context for a new approach to urban regeneration in Scotland. This new approach acknowledged that many previous regeneration initiatives had not benefited the municipal peripheral estates on the outskirts of Scottish cities, where private investment could not be attracted. The *New Life for Urban Scotland* initiative therefore sought to redirect the focus of regeneration policy away from city centres, and towards peripheral estates. The resulting designated 'Partnership' initiatives, including the Whitfield Partnership, are now complete, and interim and final evaluations have been produced for each.

Within the Partnerships designated by *New Life for Urban Scotland*, the Scottish Office was to lead specific initiatives which would apply a more co-ordinated approach to urban policy, thus avoiding the problems of fragmentation of regeneration programmes which had been recognised in the 1980s (Boyle 1990). Consequently, the Partnerships were intended to involve an integrated, strategic and multi-sectoral approach to regeneration, incorporating a significant input from the private sector (Bailey *et al* 1995). Hence economic, environmental, housing and social objectives were to be pursued simultaneously in each of the Partnership areas. The ten-year timescale proposed for each Partnership was intended to allow a strategic approach, and regeneration strategies for such areas were expected to be formulated and implemented with the clear involvement of the local community, as confirmed in the 1993 Scottish Office report *Progress in Partnership* (Scottish Office 1993a).

The rationale for such a participative approach was the perception that previous initiatives such as the *Glasgow Eastern Area Renewal* (GEAR) scheme had not adequately taken into account the views of local people, who did not feel a sense of ownership of the initiatives (Bailey *et al* 1995). The rationale for the new integrated, multi-sectoral approach was the perception that SDA initiatives, whilst having brought substantial physical improvements, had largely failed to improve employment prospects or social conditions in the areas concerned (Bailey *et al* 1995; Kintrea, McGregor, Fitzpatrick and Urquhart 1995). Consequently, physical improvements to housing areas such as Ferguslie Park in Paisley and Barrowfield in the east end of Glasgow were perceived to have failed in terms of wider regeneration aims (Kintrea *et al* 1995).

The *New Life for Urban Scotland* statement therefore proposed that the following key principles should underpin urban renewal:

- Economic improvement;
- A Comprehensive approach;
- Avoidance of 'quick fix' solutions which did not tackle the deep-rooted nature of problems;
- Involvement of local people; and
- Involvement of the private sector.
  (Cambridge Policy Consultants 1999).

It is important to note that this was presented as an experimental approach. The Partnerships were designated in four peripheral estates of different sizes throughout Scotland, as follows: Wester Hailes, pop. 12,000, (Edinburgh); Whitfield, pop. 6,000, (Dundee); Castlemilk, pop. 18,000, (Glasgow); and Ferguslie Park, pop 5,000, (Paisley). In addition to containing a large amount of housing in poor condition, each of these areas was characterized by low incomes, high unemployment, low skill levels, low educational attainment, a degraded environment, inadequate shopping and community facilities, and high levels of crime (Scottish Office 1990). Following the completion of these four initiatives, final evaluations were conducted to assess the extent to which they have achieved their objectives. The experience of the Whitfield Partnership is considered below.

The Whitfield area

Whitfield is a peripheral housing estate on the edge of Dundee, and it was designated as a Partnership because of its specific problems of housing and social deprivation. The estate is on the outskirts of Dundee, and in 1988 much of its housing was extremely unsatisfactory because of structural problems and design faults, which led to high vacancy rates, sometimes approaching 60–70% (Whitfield Partnership 1989). The estate had originally been planned to house around 12,500 residents in a housing stock comprising 4,400 dwellings. It consisted in 1988 almost entirely of local authority housing stock, with 94% of this owned by Dundee District Council (as was). However, there had been a limited amount of development in terms of owner-occupation and housing association activity, with 5% of the stock privately owned and 1% owned by social landlords (Cambridge Policy Consultants 1999). In addition, the estate had been subject to rapid population decline as well as high turnover rates and a high proportion of unoccupied units. The population of Whitfield was 9,000 in 1988 (with a total city population of 174,000 at this time), and the area lost 50% of its population from 1981-89.

The northern part of the estate was known as Whitfield Skarne. The design of this area was based on the Swedish Skarne system, popular in the 1960s. Around 2,500 homes had been built here in 1971, including 135 deck-access blocks of 4 and 5 storeys, with the original idea deriving from the aim of each resident having continuous covered access from their front door to a shopping area. Each of these blocks contained around 18 houses on average, and these were system-built using concrete in the form of 11 crescents or linked hexagons. In the southern section of the estate, known as Lower Whitfield, built from 1968–75, there was around the same number of houses but in the form of flats and cottage-type houses as well as deck-access maisonettes, extending across a landscaped open space of around 37 acres. At the western area of Lower Whitfield there were four multi-storey blocks providing 360 flats, built in 1969 (Whitfield Partnership 1988). While the condition of the housing varied, many of the deck-access blocks were in particularly poor condition. Specifically, the layout of the housing blocks led to a series of problems including physical defects such as dampness, draughts and mould, and door and window systems were also poor (Cambridge Policy Consultants 1999). In 1988, 94% of the housing in the area was local authority-owned (as compared to 48% for Dundee as a whole.

The housing problems of Whitfield derived largely from the flawed design of the estate. For instance, the deck-access design did not allow residents to control behaviour on the landings or indeed elsewhere within the vicinity. Thus the quality of housing was worst in the deck-access flats of Skarne and Lower Whitfield. This was borne out by the fact that tenancy turnover and voids were highest in the deck-access blocks. The design also led to problems of lack of privacy and security, since there was a lack of clear definition between public and private open space. Moreover, the layout of the large crescents was confusing, with large unused parking areas and rows of garages. A further problem derived from the large numbers of voids and derelict properties. In addition, there was a significant lack of diversity in the population mix, since for instance the small amount of owner-occupied housing in 1988 was all situated at the northern end of the estate (Whitfield Partnership 1988). As a consequence of such factors the area suffered from a particularly poor image, which compounded the decline in population, and 1,000 of the 4,400 units became vacant, most of these being in the Skarne blocks.

Social problems in the area were also high, since the area had a particularly transient population. Household incomes were also low; in 1989 it was estimated that 87% of households had an annual income of £10,000 or less, with 70% of residents having an annual incomes of 5,000 or less, and 35% of residents having annual incomes of less than 2,500 (Kintrea *et al* 1995). Moreover, there was a particularly high level of dependence on income support and housing benefit. Similarly, unemployment in the estate ran at twice the Dundee average, and in 1988 it was estimated that at least 40% of the men in the area were unemployed, representing 75% of the total Whitfield residents unemployed. In addition, around 350 women from Whitfield were on the unemployment register, with the highest rates being amongst those in their early 20s and in their 40s. Furthermore, 20% of households in Whitfield were lone parent households (compared with 6% for the area as a whole), and recorded crime was high, as 242 crimes per 1000 (compared to 123 for Dundee as a whole).

The Whitfield Partnership

Factors such as the history of previous initiatives in the area, and the previous prioritization of the estate by many of the partners, supported the selection of Whitfield as a Partnership. For instance, the 'difficult to let' nature of the estate had led to physical improvements by Dundee District Council (DDC) such as removal of some of the deck-access blocks, and the estate had been targeted for housing renewal by DDC at the start of the Partnership. Moreover, a policy of tenure diversification had been followed in the area before the Partnership, and DDC had been attempting to bring about transfers of housing stock out of the public sector since 1984, because of the need for additional investment in the area. In addition, Tayside Regional Council (TRC) had developed a social strategy for the Region in 1987, which included Whitfield as a target area. However, such initiatives had been implemented in a relatively unco-ordinated and incremental fashion. Consequently, the Whitfield Partnership set out to bring about a co-ordinated strategy and investment programme that would ensure that change was progressed in terms of housing, employment and training, and community facilities.

The Whitfield Partnership was established in June 1988 so as 'to develop and co-ordinate the implementation of, in Partnership, a strategy for the long-term regeneration of Whitfield in line with the government's objectives for urban regeneration as set out in the White Paper' (Whitfield Partnership 1988, 1). The Partnership proposed 'a broad-based strategy rather than a detailed "blue-print"' (Whitfield Partnership 1988, 1), and it was explicitly intended to build on the strengths of the area as well as on the initiatives that had already been taken to improve the area.

The structure of the Whitfield Partnership was similar to that of the other Partnerships, and the Partnership Board was chaired by a senior official of the Scottish Office. The Board comprised members from Dundee District Council, Tayside Regional Council, Scottish Homes, Scottish Enterprise Tayside, Tayside Police, the Employment Service, Dundee and District Chamber of Commerce, Dundee Enterprise Trust, Tayside Regional Health Board and the Whitfield Business Support Group. The residents of the estate were also represented by the Whitfield Steering Group, and the Board was serviced by the Partnership Strategy Team, which was located in Whitfield. This team was small, initially comprising two Scottish Office staff, since it was assumed that the delivery of the Whitfield Partnership Strategy would be undertaken by the individual partners. This appears to have avoided the problem experienced in other Partnerships such as Wester Hailes of the equivalent team being seen to lead too strongly, and the Whitfield team was careful to avoid taking responsibility away from the other partners (Bailey *et al* 1995).

The physical set-up of the Whitfield Partnership was unlike that in the other Partnership areas since no special accommodation was built to house it. Instead, it was located in several Portacabins, which later came to be used by the Job Placement

Service and the Training for Jobs Initiative (Cambridge Policy Consultants 1999). Sub-groups of the Partnership Board were also set up to assist the development and implementation of policy, and these covered the areas of housing and environment, community services, employment and training and monitoring and evaluation. These four sub-groups usually provided the forum where detailed issues were discussed.

Objectives

The *First Report on Strategy* (Whitfield Partnership 1988) indicated that employment, housing and community services were the main priorities to be addressed, and a series of objectives in relation to the area was developed, focusing around short-term, tangible outputs. Significantly, poverty was not identified as a specific issue for the Partnership to address, and the Strategy Report indicated that it was anticipated to take 5-10 years before significant achievements could be made. The main objectives set out in the First Report are considered below.

Housing

It was immediately clear to the Whitfield Partnership the whole of the estate needed physical upgrading to some extent, with the deck access housing in Skarne and Lower Whitfield in most urgent need of upgrading. The *First Report on Strategy* considered a series of immediate objectives. First, the priority in this context was the objective of improving the quality of housing, the immediate environment and resident's privacy and security. This was aimed to enable residents to lead normal lives and was seen to entail the need for a radical re-planning of the whole environment in terms of deck-access housing in particular, as well as more effective local management and maintenance activity.

Second, there was the objective of developing greater resident control of the housing and commitment to the area. This was aimed at encouraging residents to become involved in bringing about improvements which would change the image of the estate radically. The involvement of residents was seen as important to ensure collective responsibility for the setting and maintenance of environmental standards, and it was recognised that tenants' associations had an increasing role to play in this respect.

Third, there was the objective of increasing the range of tenures and house types in the area. In this context it was recognised that the scale of dereliction in part of the Whitfield estate was such that there was no single answer in terms of the offering of new choices of tenure to residents, or the large-scale attraction of new residents into the area. This was partly because it was anticipated that many local residents would not wish to contemplate other forms of tenure. However, it was also expected that alternative forms of tenure would become available, for instance in terms of housing associations, community ownership and 'right-to-buys' (following from legislation allowing tenants in local authority accommodation to buy their homes from the local authority). In addition, it was anticipated that some new residents could be attracted into the area, at the same time as the quality of housing was improved and new initiatives were undertaken in terms of tenure choice, and that there would be a gradual progression towards a higher percentage of owner-occupation.

Fourth, there was the objective of reducing the rate of turnover and achieving a stable community. This was judged to be a separate issue from the increase in tenure choice since a more appropriate mix of household composition was also seen as necessary. It was recognised that maintaining the area's population at anywhere close to its 1988 population of 3,500 would depend on factors separate to the housing stock, particularly in terms of economic considerations and resident control. Fifth, there was the objective of devising mechanisms for co-ordinating, developing and diversifying housing provision through Dundee District Council as well as other agencies and the private sector. The essential objective in this respect was to ensure that the different statutory and funding responsibilities of the relevant agencies were effectively meshed so as avoid unnecessary overlap of functions (Whitfield Partnership 1988).

The Partnership thus sought to develop a clear strategy for housing which, as indicated above, had previously only been addressed in an incremental fashion. Soon after designation of the Partnership, specific targets for improvement were established, such as to reduce the number of vacant houses by 50% by the end of 1991. A wider housing strategy was set out in the Partnership's 1990 *Strategy Report*, and in 1991 Dundee District Council (DDC) and Scottish Homes signed a Strategic Agreement which identified Whitfield as a top priority area for capital spending. The overall approach to housing was therefore development-led, with little attention being given to management issues, and demolition was a major element of this strategy because of the high costs of refurbishment. This represented a change in direction for the local authority, which had originally anticipated a strategy relying on improvement (Kintrea *et al* 1995).

The environment

The Whitfield estate was physically separated from the rest of the city by the Dighty Valley. The housing was set in an open plan environment, with open space at the centre of the area intended to improve pedestrian movement by excluding cars, and a network of roads and paths designed to separate vehicles from pedestrians. However, there were major problems associated with the environment in the area. For instance, the predominantly east-west alignment of the main road routes in the area ran counter to the network needed to integrate Whitfield into the structure of the city. Moreover, the separation of vehicles and pedestrians in practice meant that the footpaths were not well used, so that tenants felt insecure, and this problem was exacerbated by the long distances between some of the residential areas and the shopping centre.

In addition, the large unused parking areas and rows of garages in the Whitfield Skarne area produced a dismal environment. The environment of the shopping centre was also poor, and little attempt had been made to integrate it with the adjacent community facilities. While there was a great deal of public open space in Whitfield, much of it was unused, particularly the central open space, which suffered from drainage problems, and the design of the estate did not allow the residents to have their own gardens. Moreover, the children's play space equipment was judged to be in the wrong place and largely irrelevant to local needs. There were also several prominent 'eyesore' sites, as well as a distinct lack of 'landmarks' (Whitfield Partnership 1988).

Because of such problems, the Whitfield Partnership established a series of detailed objectives in relation to the environment. First, there was the objective of integrating Whitfield more fully into the structure or the city as a whole; this was to be achieved by improving the radial road system as well a footpaths. Second, there was the objective of improving accessibility within Whitfield, by means of redressing the inherent design faults in this respect and improving aspects such as sight lines on footpaths. Third, there was the objective of creating a better residential environment by means of improving the identity of sub-areas, providing more defensible space, and reducing the size of large areas of garaging; this was identified as the priority objective in terms of the environment within the First Report (Whitfield Partnership 1988). Fourth, there was the objective of improving the general environment by means of planting for instance.

Employment and training

In addition to the problems of unemployment rates highlighted above, Whitfield suffered from the lack of any major industrial or commercial employers in the area. Furthermore, the only Job Centre available was in the centre of the city. It was recognised that economic activity in the Whitfield area would have to be developed in the context of the policies and activities of the Dundee Project, and several specific objectives were developed in this context. First, there was the objective of creating new opportunities for employment for Whitfield residents, including the development of small businesses, and the Whitfield Partnership anticipated the creation of a 'Whitfield Employment Initiative' which could pursue such employment and training objectives. Second, there was the objective of developing residents' skills and improving their ability to compete for jobs, with particular emphasis on women wishing to return to work. Third, there was the objective of devising new training programmes to meet local needs. Fourth, there was the objective of providing small workplaces locally or readily accessible to Whitfield residents; this links clearly to the aim of building new opportunities for small business formation. Fifth, there was the objective of improving provision in the area of information about job vacancies. Sixth, there was the objective of targeting specific sectors of Whitfield unemployed residents for training, including, again, women who wish to work (Whitfield Partnership 1988).

Community facilities and quality of life

The Whitfield Partnership recognised that new and improved services were needed in the area in order to achieve social objectives. The requirements included more training and information, more recreational provision, improved education, and more appropriate shopping provision. Consequently, it was intended to make more effective community use of the facilities at Whitfield High School, the Whitfield

Activity Complex and the Greenfield Community Centre, as well as extending the range of neighbourhood community lounges, based on existing facilities, provided a valuable focal point for residents and young children to meet. It was also intended to investigate the possibility of providing new indoor sports facilities. The need to maximise the activities of existing voluntary organizations was also recognized, as well as the need to provide a range of information and communications services. It was also seen as necessary to promote services for children and young people, and to improve the range and quality of health care provision. Moreover, it was intended to make better use of existing Urban Aid funding, to promote the 'Safer Cities' project in Whitfield and to examine options for improving shopping provision (Whitfield Partnership 1988).

Achievements

Housing

Most of the spending undertaken by the Partnership was aimed at improving the quality of the housing as the physical environment, for reasons that are set out above. Because of the fundamental design flaws of the estate as a whole (as set out above) it was decided that the most cost-effective way forward was to demolish many of the existing housing blocks. This part of the strategy was achieved ahead of schedule, and by 1994 over 1,500 houses were demolished, 520 houses were comprehensively improved, and around 1,600 houses benefited from some improvement (Kintrea *et al* 1995). By 1995, around 2,000 dwellings had been demolished, many of which were deck-access, and a further 200 demolitions had taken place by 1998. In total around half the stock was demolished, a figure that was higher than in any other Partnership area.

In addition to housing demolition, 650 new build housing units were constructed, of which 390 were built by three locally-based housing associations, with 260 arising from private developments of owner-occupied housing. Furthermore, 2,600 properties were improved. Overall, the replacement of demolished stock was only partial since the density of the new units was lower than that of those that had been demolished. In terms of gross public expenditure on housing renewal, £39.4 million was spent in Whitfield from 1988-98. There was also some development without subsidy, comprising the former Aberlady site, and every £100,000 of public expenditure levered in £21,000 (Cambridge Policy Consultants 1999).

In terms of additionality, it is estimated that the average additionality ratio for housing renewal in Whitfield was 68%, compared to an average for the four housing partnership areas of 63%. In terms of outputs, from 1988-98, 2,603 units were improved or rehabilitated in Whitfield, out of 4,400 units in total, with 2,417 demolitions. There were also 663 new build units, with a net change in stock of -4%, compared to the average decrease across the four housing partnerships of -3%. In addition, housing tenure was diversified to a significant extent, with a fall in the share of local authority ownership from 94% in 1988 to 44% in 1999. There was also an increase in the share of owner-occupied housing from 5% in 1988 to 41% in

1999 (Cambridge Policy Consultants 1999). In terms of outcomes, the proportion of residents who were very or fairly satisfied with their accommodation rose from 77% in 1988 to 85% in 1998, and the proportion of residents who were very dissatisfied with their accommodation fell from 9% in 1988 to 4% in 1998.

Again, however, population flow must be taken into account since only 53% of the households in Whitfield in 1998 had been resident before 1988. Moreover, an important point in this context is that the type of tenure provision has significant implications for other crucial aspects of regeneration; for instance, new private or housing association units are likely to be unavailable to the most deprived. This may indicate that in Whitfield problems were 'exported' rather than solved. However, action taken to diversify tenure clearly slowed the rate of population decline, and Whitfield's increase in owner-occupied stock would seem to be linked to the marked increase in resident employment in the area.

The environment

Many of the objectives in relation to the environment were addressed by means of the radical improvement programme in relation to housing, particularly the demolition and re-build programme. In addition, a number of projects were implemented to improve the overall environment on the estate, including work on footpaths, lighting, car parking, fencing and walls, clothes drying and refuse facilities, play and landscaped areas and anti-vandalism features (Cambridge Policy Consultants 1999).

Employment and training

There were many achievements in relation to objectives for employment. For instance, a job placement service that proved to be extremely successful. There were also a number of training initiatives, including the Whitfield Learning Shop whose activities continued in the form of the Whitfield Learning Centre. IT and associated skills were also encouraged, and Workstart, a training provider which later changed its name to WS, was encouraged to provide a variety of training courses for Whitfield residents in such areas as woodwork, metalwork and fork-lift truck driving (Cambridge Policy Consultants 1999). WS Training was established as the Claverhouse Group in 1996, which is now a city-wide provider that targets the poorest 20% Enumeration Districts, including Whitfield. The Partnership also provided advice to new businesses and supported the Community Economic Development Whitfield Group, a community business employing eight staff which undertook a range of community projects. There was also a limited amount of employment geared to education as part of the Partnership's training and employment initiatives, though this mainly took the form of co-operating with city-wide initiatives (Cambridge Policy Consultants 1999).

Again, however, it is important in this context to stress the linkages between employment and housing, and between employment and the quality of life experienced by the residents of Whitfield (Cambridge Policy Consultants 1999). This is because many of the employment measures set up in the area, in common with those in other housing partnership areas, were predicated on the presumption of a relatively stable population. Hence it was assumed that the problems arising from lack of employment would be gradually reduced over time. In practice, however, there was a large amount of 'churning' in the population, with people moving into and out of the estate. There was also 'churning' in the labour market, with people finding and quickly leaving work which was seen to be low-paid and with relatively little prospects. This means that the problem of local unemployment was continually exacerbated.

However, in terms of activities relating to job opportunities and access to employment, it is estimated that the Whitfield Partnership, relative to the three other housing Partnerships, was relatively strong in terms of encouraging greater access to the labour market. For instance, a Jobcentre was located in the estate in 1989, and the Partnership developed innovative approaches to addressing the barriers to employment that faced disadvantaged groups, including the long term unemployed, women returnees and youth. It also provided innovative training initiatives, exploited opportunities for new employment in peripheral industrial estates, and made a valuable contribution to community businesses in Whitfield. In addition, two particularly good examples of practice in Whitfield may be identified. First, the Partnership provided a range of pre-employment and vocational training options; second, the inclusion of two additional development workers from Scottish Enterprise Tayside enabled greater outreach to people who directly understood the needs of the unemployed client group (Cambridge Policy Consultants 1999).

However, the Partnership's activities were les successful in some respects. In particular, there was only moderate business support, with one of the board members being provided by the Business Support Group. In addition, only a small amount of employment was created in terms of the leisure and retail sectors (though the shopping centre was eventually refurbished).

Community facilities and quality of life

In terms of community services, the Partnership had moderate success for instance in terms of the completion of the Whitfield Health Clinic, which enabled new and improved services to be provided including midwifery and antenatal care, an eye clinic, speech therapy and a chiropody clinic. The Partnership also supported the Health Steering Group and the Clinic Users Group, and it increased the provision of childcare facilities including the Whitfield Community Nursery (Cambridge Policy Consultants 1999).

In terms of leisure facilities, the Partnership put considerable time and resources into a possible local sports facility. This was not successful, though a multi-sports facility for children and 7-a-side football pitches were developed on the central open space. The Partnership also promoted the environmental improvement of the shopping centre and non-retail premises, since this open air shopping mall in 1980 offered only a few shops selling convenience goods, with little choice and high prices. The Partnership attempted to encourage non-retail uses, with the result of the Whitfield Health and Information Project, described below. It also sought to encourage development of the shopping centre, and purchase of the site by a developer eventually resulted in refurbishment works that levered in monies from the European Regional Development Fund, Scottish Enterprise Tayside and Scottish Homes. The shopping centre was re-launched in 1998, but there were concerns that the development did not provide a critical mass, given the presence in the city of competing sites for potential occupiers (Cambridge Policy Consultants 1999).

In terms of wider quality of life factors, there was a significant improvement in levels of poverty on the estate, with a fall in the proportion of residents with incomes under £100 per week from 64% in 1988 to 48% in 1998 (Cambridge Policy Consultants 1999). In addition, crime is an important issue in this context it was of central concern to residents. In 1988, 83% of residents did not feel that the police made sufficient efforts to meet local people, and the area had a higher incidence of crime than the city as a whole. Partly as a consequence, a new police station was opened in the area in the late 1980s. In addition, Whitfield played a prominent part in the *Safer Cities* Scottish Office initiative of 1990, which led to projects concerned with safety in the home, womens' safety (self defence), victim support and personal attack alarms. Primary schools were also given crime prevention and safety instruction. Moreover, the attitude of residents to the police appeared to change as a result of the introduction of community policing to Whitfield. However, while levels of crime decreased, the fear of being out alone after dark increased, with 42% of residents affected in 1998 and 63% affected in 1998. Some of this fear may be explained by the greater incidence of drug use on the estate in 1998 compared with 1988. Nevertheless, overall levels of satisfaction expressed by residents on the estate increased significantly during this period, with only 5% of residents very satisfied in 1988 as compared with 25% in 1998 (Cambridge Policy Consultants 1999).

In terms of education, the main focus in Whitfield was on pre-school initiatives, youth and adult education, and childcare, with the establishment of the award-winning Whitfield Childcare Centre. In terms of youth provision, it is unfortunate that the proposal to build a sports facility failed because of the limited market and high capital and running costs. However, a scheme was also established to subsidise the use of leisure and sports facilities by local schoolchildren during the school holidays. In terms of health, 80% of residents were at least 290 minutes away from a GP, though the Whitfield Health and Information Project was supplemented by the development of a fully functioning Health Centre in 1993 This included provision of a drop-in centre, a library, facilities for self-help groups, womens' groups for fitness, weight loss and healthy babies, a healthy eating project, a drugs information project and an alcohol abuse project (Cambridge Policy Consultants 1999).

Overall evaluation of progress

Having set out the main objectives and achievements in relation separate thematic aspects of the work of the Whitfield Partnership it is helpful now to consider the overall effectiveness of the Partnership and the factors that contributed to it. These are set out below in terms of the methods of evaluation, the development of a regeneration strategy, effectiveness of spending overall, community participation and partnership working, and arrangements for succession.

Evaluation methods

In order to facilitate effective evaluation of the housing partnerships, baseline studies were undertaken for each area, with the Whitfield Baseline Report being published in Spring 1989. This set out the starting points in terms of the main problems in the area, as outlined above. It also provided a detailed statement of social and demographic characteristics, economic and employment factors, housing condition and satisfaction characteristics, community facilities and services, leisure facilities, and levels of crime in the area (Whitfield Partnership 1989). It therefore set out the main features to be taken into account in developing a strategy for the area. A household survey was also undertaken in 1988, and the household survey was repeated in 1994 when all four of the housing partnerships were the subject of separate interim evaluations. A final evaluation was prepared in 1999.

Development of a regeneration strategy

From the outset, a key element of the partnership approach to be adopted was the development of a strategy for regeneration produced by the Partnership with a set of clear objectives agreed by all of the local partners. Broadly, attention was given mainly to housing improvements in the early years of the strategy. This was because it was clear that this was a means by which visible outcomes could be achieved so as to secure the commitment of all the partners involved as well as the wider credibility of the Partnership (Cambridge Policy Consultants 1999).

The first act of the Partnership was to undertake a thorough assessment of the nature of the problems that existed at the outset as well as of the impact of the programmes that were already in place. Hence the *First Report on Strategy* was presented to the Secretary of State for Scotland in December 1988. This Report placed particular emphasis on the need for an overall plan for housing, a Whitfield Employment Initiative and a review of the provision of community services. These priorities subsequently formed the rationale for the creation of the three operational sub-groups of the Partnership. In April 1990, the Whitfield Strategy Report was published, which set out the priorities for policy together with a clear strategy for implementing the policies in co-ordinated manner, with clear aims, objectives and targets. This strategy formed an initial plan which could be used to monitor subsequent activity (Cambridge Policy Consultants 1999).

Building on the analysis in the *First Report on Strategy*, the Strategy Report set out its priorities as follows:

a. Housing and Environment:
    *   To improve housing quality;
    *   To reduce turnover rates, stabilise the community and attract people onto the estate;
    *   To increase tenure choice; and
    *   To improve the quality of the environment.

b. Employment and training:
- To reduce unemployment to the Dundee average;
- To enable residents to acquire new skills;
- To support those who wished to start their own businesses; and
- To encourage women who wished to return to work.

c. Community services:
- To allow residents to gain access to a range of social, leisure an recreation facilities;
- To improve levels of health care;
- To increase child care provision; and
- To increase levels of security and reduce levels of crime.

These priorities can be seen to have been derived from the analysis of problems in the area contained in the 1988 *First Report on Strategy*. They clearly represent an attempt to satisfy the remit common to all the housing Partnerships, namely to bring about a comprehensive approach to social, economic and environmental regeneration (Scottish Office 1993b). However, it is clear that no new spatial plan was formulated as a result of the strategy, since many of the components had been initiated by Dundee District Council (as was) prior to the Partnership. Moreover, no new partners were brought into the strategy as a direct result of the Partnership, since all the participants were either involved prior to Partnership designation, or had been identified before the strategy was worked out (Kintrea *et al* 1995).

Effectiveness of spending

In terms of expenditure, approximately 2% of the total spend of around £55.5 million from 1988-1994 was needed to run the Partnership office, about 4% was used for employment and training schemes, and about 8% for community services. The bulk of spending, however, was on the housing and environmental programmes, which together accounted for about 85% of total spending (Kintrea *et al* 1995). Only a small amount of private sector investment was attracted, largely for house building and loans to housing associations, and the leverage ratio between public and private spending was around 6:1 (Kintrea *et al* 1995). In terms of overall public expenditure within the Whitfield Partnership area from 1988 to 1998, £2.8 million was spent, the bulk (£2 million) of which was spent on training, including all forms from pre-employment to customised skill training. In terms of the source of expenditure, most of the spending in Whitfield was spent by Scottish Enterprise Tayside (£1.4 million). However, the total spending per head of the 1988/90 population of working age in Whitfield, amounting to £1,250, was the lowest figure of any of the Housing Partnerships. This was in part the result of a conscious decision to channel spending in Whitfield through the existing organizations.

In terms of additionality of spending in Whitfield, it is significant that Dundee District Council and Tayside Regional Council had put funds into a range of projects aimed at improving circumstances in Whitfield, prior to the *New life for Urban Scotland* policy statement. While the spending in the 1980s was relatively low, over £400,000 of expenditure was committed in 1988-89, immediately before the establishment of the Whitfield Partnership, and this money was earmarked for new projects designed to address local problems of deprivation. The expenditure increased significantly as a result of the Whitfield Partnership. Moreover, in 1989 the general level of spending was being continued, partly because of previous commitments made by the Partnership, long after the formal closure of the Partnership in 1995-96 with the withdrawal of the Scottish Office (Cambridge Policy Consultants 1999).

In terms of advice and guidance and training activity, the levels of additionality were judged to be the highest of the four Housing Partnerships. Moreover it is significant that each of the four Housing Partnerships had different priorities in terms of their intervention in the local labour market, with Whitfield placing particular emphasis on advice, guidance and counselling. As a consequence, the Whitfield Partnership was successful in generating 2,042 extra job placements, though it only contributed 551 additional training places for residents.

In terms of outcomes achieved, the percentage employment rate improved sharply in Whitfield, from 42% in 1988 to 62% in 1998 (Cambridge Policy Consultants 1999). This represents an important indication of comprehensive regeneration since employment reduces dependency on state benefits as well as reducing poverty. Again, however, the changing population needs to be taken into account in this context since there was an outward flow of 33% of the population between 1988 and 1998. While this may be explained in part by forced outflows due to housing demolition, a contributory factor arises from the presence of programme beneficiaries who are enabled to leave the area through choice. Hence it may be argued that the labour market problem in Whitfield was recreating itself. Furthermore, the positive outcomes in Whitfield may be explained in part by the fact that those decanted out due to demolition or refurbishment are likely to be relatively disadvantaged, with new or refurbished stock being subsequently filled by employed people. This may indicate therefore that some problems have simply been 'exported' from the area, as suggested previously.

Nevertheless, it is estimated that in the absence of the Whitfield Partnership, some £200,000-£300,000 worth of annual expenditure would have been likely. Hence it may be concluded that almost three-quarters of the spending which did occur during the period 1988–98 was as a direct consequence of the Partnership (Cambridge Policy Consultants 1999).

Community participation and partnership working

The participative approach of the four housing Partnerships was intended as an alternative to the GEAR model followed in the 1970s, which was seen to have marginalised the role of the local population to an extent (Kintrea *et al* 1995). Hence the Scottish Office emphasised the importance of community participation in the work of the Partnerships in its 1993 *Progress in Partnership* report. Specifically, this states that 'the experience of the Partnership approach has confirmed the wisdom of involving local communities fully in regeneration plans for their areas' (Scottish Office 1993, 24). However, in terms of the Partnerships as a whole, the experience of local communities was ambivalent and sometimes unproductive (McArthur *et al* 1994).

The broad arrangements of the Whitfield Partnership in terms of partnership working are set out above. In fact the District Council had begun to work with local tenants' action groups before designation of the Partnership, particularly in terms of housing improvements. Initially, only four community representatives were entitled to attend the meetings of the Partnership Board, but it was argued by the steering groups that this did not give the community an important enough voice in the workings of the Partnership. Consequently, it was decided to allow ten community representatives to attend, and four members from the community services sub-group were allowed to sit on each of the Partnership's other sub-groups (Cambridge Policy Consultants 1999).

After 1988, an umbrella group of representatives known as the Whitfield Community Steering Group was set up, with 20 members drawn from seven designated areas of the estate. As indicated above, this group had a maximum of ten representatives on the Partnership Board and four representatives on each of the sub-groups. However, because many residents felt that they were not adequately represented, the original model based on geographical representation was dropped, and replaced by a system that allowed any community organization to become a member of the Steering Group and to send two representatives to it (Cambridge Policy Consultants 1999).

Overall, the level of involvement by the community in the Whitfield Partnership was varied, and the feeling was frequently expressed that community involvement was something of a token gesture (McArthur *et al* 1994). Whilst the community's contribution to the implementation of housing and environmental improvements had been significant, it was widely felt that its overall impact on the development of policies and programmes had been limited (Kintrea *et al* 1995).

In terms of the private sector as partners, the Whitfield Business Support Group was set up in 1989 to enable the private sector to become involved in the work of the Partnership. It identified employment and training as the area where it could best make its contribution to the Partnership, and it therefore operated mainly through the employment and training sub-group of the Partnership. Partly as a result of its efforts, unemployment in Whitfield fell by 41% from 1988–1991, as compared to a 19% fall for the city as a whole (Kintrea *et al* 1995). However, this success was not matched by its contribution to other aspects of the Partnership's work, and the main source of private sector involvement in the Partnership was in terms of private house building (Cambridge Policy Consultants 1999).

Arrangements for succession

The arrangements for succession were important aspects of the work of the Whitfield Partnership. There was a presumption in 1992 that the targets set in the original Strategy for regeneration would be met by 1995, so the Scottish Office withdrawal from the chair of the Partnership was timed for 1995. By this time, the view of the Board was that most of the aims set out at the outset had been achieved. The role of the Partnership was seen as to create the conditions within which others could successfully operate. It would therefore act as a 'policy entrepreneur', which could build the capacity of individual players who would ultimately act to take forward the action programme of the Partnership. Specifically, the Partnership was seen as a catalyst that could identify and assess project feasibility, bring together resources and expertise, and co-ordinate the work of the implementing bodies (Cambridge Policy Consultants 1999). However, it was recognised that there was a need for a clear exit strategy as well as successor arrangements, but unfortunately these arrangements proved to be weak. This lack of a clear exit strategy would seem to be linked to the assumption that ongoing regeneration needs would be followed up as part of the (then) emergent Programme for Partnership initiative, with the result that there was a significant loss of momentum in the regeneration of the area.

In 1993, it was envisaged that the community would chair the Partnership after the withdrawal of the Scottish Office, and that the Sub-Groups would continue. After formal closure of the Partnership in 1995, the Scottish Office continued to chair the Employment and Training Sub-Group for one year, and after its total withdrawal in 1996 the Partnership Support Team was disbanded. The community did take up the chair of the Partnership, but there was a feeling that, following the withdrawal of the Scottish Office, the seniority of those attending the 'Continuation Partnership' dropped. The result was that decisions could not be made at the Partnership meetings with the same degree of certainty as before, since the representatives had to refer back decisions to other within their own organizations. Part of the reason for this lowering in seniority would seem to have arisen because of the emergence of other areas (the Priority Partnership Areas, subsequently converted to Social Inclusion Partnerships) in the city which needed to be given more attention since Whitfield had now 'had its share'. In addition, the reform of local government in 1996 would also seem to have contributed to this problem.

The withdrawal of the Scottish Office from the Whitfield Partnership provided some important lessons for transitional arrangements in the other Partnerships, since in Whitfield there were 27 changes in personnel within the partnership network. It was felt that a large part of this was due to what might be called 'partnership fatigue'. The lessons seem to be that there is a need in such circumstances to focus the community representation as well as, where necessary, to replace individual project leaders (Cambridge Policy Consultants 1999).

Recent activities

It is significant in this context that, in 1996, 30% of all spending under the Urban Programme was focused in the former Whitfield Partnership area, in which there were ten Urban Programme projects. Moreover, during the period 1988-94, around £70 million was spent in the area, focusing on physical improvement, and with a strong emphasis on housing renewal and replacement. In parallel to this activity, a strong community-led delivery structure was created. Moreover, in 2004, it was clear that the Whitfield area had, largely as a result of Housing Partnership funding in the 1980s, significantly changed its tenure pattern and achieved significant levels of investment in terms of housing and local facilities.

Nevertheless, problems remained in terms of lack of local employment opportunities and lack of local shopping (though there are good bus links to nearby facilities) and leisure facilities. Problems also remained in connection with the remaining local authority housing stock, and there is a programme of continued demolitions, with 36 houses demolished in 2005-06 (Dundee City Council 2005b). Moreover, census information indicated that the area had higher levels of unemployment (at 9.6%) than the Dundee average (at 5.4%), and that income levels in the area remain relatively low, with the area bring reported as 16th in the 100 most deprived areas in Scotland in 1993.

The regeneration of the area remains a priority of the City Council, and further new build development by housing associations is planned as part of the development plan of Communities Scotland. In addition, while the local authority continues to hold the majority of the housing stock, there is an increasing incidence of mixed tenure in the area, with evidence of private development for sale to the east and north of the area. However, the area retains a strong sense of local identity (Dundee City Council 2006).

Conclusions

The problems of the Whitfield housing scheme may be traced back to the Dobson Chapman Report of 1952, as indicated in Chapter 5. This paved the way for the development of peripheral housing areas, and development of Whitfield by means of (then) innovative and cheap construction methods parallels in many respects the now-evident errors in many similar schemes in many other cities in the UK and elsewhere. In the case of Whitfield, such problems, exacerbated, as in many other cities, by residential selection policies and practices as well as by physical aspects of design, were compounded by the physical separation of the area from the central part of the city. Of course such issues have been felt in other Scottish cities, particularly Glasgow, where public sector peripheral housing were also developed along similar lines to Whitfield but on a larger scale.

Hence Whitfield has many similarities with other areas in other UK cities in terms of problems and origins. However, the application of the Whitfield Partnership was a critical innovation, arguably building on both the Scottish tradition of pragmatism and partnership in planning and regeneration, as well as the well-developed community infrastructure within Dundee itself. Moreover, in spite of shortcomings as set out above, major improvements have clearly been achieved by the Whitfield Partnership, for instance in terms of housing condition, tenure diversification and enhancing the employability of residents. In addition to concrete outcomes, it is clear that the perceptions of the estate held by those outside the area have also been improved. However, the provision of shopping and leisure facilities has been disappointing, and the wider input of the private sector into the formulation of strategies has not been significant, though this is increasing. Nevertheless, perhaps the most important aspect of progress in terms of the key lessons arising from the Whitfield Partnership has been the clear value of the mode of partnership working adopted. This has brought a number of advantages; in particular, the Partnership acted as a central organization that could confer credibility and legitimacy, and the

Board and Sub-Group structure allowed effective co-ordination by means of joint projects and joint funding.

Of course, an important aspect of the Partnership's success is the extent to which the outcomes would have been obtained without the designation of the Partnership. In Whitfield, the tangible physical improvements to housing are the most evident effects of the Partnership, though it may be suggested that many of these improvements could have been achieved in the absence of the formal Partnership mechanism (Kintrea *et al* 1995). However, the final evaluation of the four Housing Partnerships indicated that the Whitfield Partnership had contributed an important aspect of good practice in terms of the use of the 'Workstart' initiative, which focused on the barriers to employment faced by residents and adopted an innovative approach to providing a range of services that proved to be extremely valuable.

Underlying all the analysis of indicators of effectiveness, however, is the fundamental issue of population change. The view of some observers that the Partnership would simply lead to the displacement of economic and social problems to other areas did not seem to have borne out by 1995 (Kintrea *et al* 1995). However, taking into account the whole of the period 1988-98, the problems of population movement and 'churning', as described above, are extremely significant. Essentially, many of the improvements would seem to have been brought about by gentrification. The lessons would seem to be that a regeneration scheme is likely to be most effective when it can retain its population, since it can then alleviate the various sources of deprivation on an incremental basis. This was not possible in Whitfield for reasons set out above.

Hence, overall it seems clear that the Whitfield Partnership initiative has provided significant benefits to the city. However, it is also clear that, as an area-based initiative, it had to some extent the effect of displacing families and individuals with social problems into other areas (Fernie 2000). Moreover, such an area-based initiative could in any event only address a small part of the wider set of social and economic problems in the city. Nevertheless, relative to analogous initiatives in other contexts, the Whitfield Partnership brought a flexible approach to strategic regeneration, allowing all local interests to be involved and to contribute in some way to both the development and the implementation of policy. This may be compared for instance with the use in England in the 1980s of Urban Development Corporations, which offered much less potential for the contribution of a range of interested parties (Bailey *et al* 1995), and which also failed in many cases to bring about a strategic, multi-sectoral approach to regeneration (Imrie and Thomas 1993).

The conclusion would seem to be that area-based initiatives such as the Whitfield Partnership can deliver benefits, but they must be situated in the broader strategic context of the city as a whole in order to avoid problems such as leakage of benefits and displacement of problems. This reflects to a degree the contemporary focus on city-region approaches to regeneration and wider policy development, as well as the experience of area-based regeneration initiatives in other contexts. Moreover, it may be argued that the Whitfield Partnership was in practice focused on physical, as opposed to social and economic, aspects of regeneration, as reflected in the outcomes achieved. The next Chapter considers those initiatives which have attempted to apply a more holistic approach to regeneration, incorporating social as well as economic and physical benefits.

Chapter 8

Holistic Regeneration Approaches

Introduction

This Chapter sets out the experience of holistic regeneration approaches in Dundee in terms of the experience of the Dundee Partnership, which applies a broadly-based approach to regeneration, and the Social Inclusion Partnerships (SIPs) initiative (as well as the PPA initiative which preceded it), which aimed to translate the vision of the Dundee Partnership into practice in terms of the areas most affected by social, economic and environmental disadvantage. It also considers the transition to the Community Regeneration Fund. In addition, it considers the role of the City Vision (following the Cities Review) and Community Plan. Finally, it evaluates the extent to which these approaches have provided a sustainable solution to entrenched urban problems.

The Dundee Partnership

The Dundee Partnership is pivotal to the changes orchestrated in Dundee, and it has matured as a form of urban regime in the specific context of the Dundee economy. In many ways, the Dundee Partnership represents the circumstances, ideas and themes prevailing at different times in the context of urban regeneration. These influences have been national in character, for instance in relation to the neo-liberal assertion of a more significant role for business in local development; and locality-specific, as shown by the way in which Dundee has responded to its particular circumstances. The Dundee Partnership may be considered a 'virtual' organization, since it does not a have an identifiable base, space or building, nor does it have dedicated staff. Nevertheless, it has a strong identity in Dundee, and has exerted considerable influence in addressing the regeneration priorities of the city.

The Dundee Partnership is an informal institutional arrangement which binds the principal stakeholders in Dundee in a coalition which works to an agreed agenda of common priorities, and it has attempted to bring about a consensual approach by means of broad agreement between the individual partners in terms of the regeneration of the city. While the Dundee Partnership is public sector-led, there are clear links to the business community, both formally and informally, via Scottish Enterprise Tayside (which is itself a private company).

The origins of the Dundee Partnership are bound up with the post-war economic history of Dundee, when the city's relative economic under-performance was acknowledged by the Westminster governments. The conventional response was the 'top-down' provision of regional industrial policy assistance applied to attract new inward investment to Dundee through incentives in the form of capital and property subsidies. It was also intended at this time to encourage Dundee firms to expand and diversify where possible, perhaps through the adoption of new technological processes. Regional industrial policy was administered from the centre, and, at different periods throughout the post war period it was applied at different levels of financial support and at different scales of geographical coverage. Essentially, Dundee was competing against other depressed regional economies, which were also experiencing relative economic depression. However, regional economic support to industry was transformed in the late 20th century. In particular, the level of support from Westminster declined as political arguments suggested there was greater scope for alternative approaches, such as those associated with the neo-liberal phase identified by Giddens (outlined in Chapter 2). In addition, European funding was applied to the city, by means of the application of 'Objective 2' status.

Within this policy framework, several local parallel actions were initiated, with the involvement of a range of local institutions, to address the need for broadly-based regeneration, as indicated elsewhere in this book. Specific problems included poor and sub-standard housing, redundant land and buildings, declining service provision and the need to provide appropriate facilities for the changing population of the city. In addition, land use planning interventions encouraged the provision of new public sector housing in peripheral housing estates throughout the 1950s and 1960s (as set out in the preceding chapter), and there was a focus of attention on the redevelopment of the Dundee's city centre in the 1970s.

The Dundee Project

The Dundee Project was an early local initiative, and it involved a partnership of the regional and district authorities and the Scottish Development Agency. It represented an important step forward in establishing an explicitly co-ordinated approach to regeneration in Dundee, though it acknowledged that individual bodies each had their own technical agendas, budgets and priorities. Moreover, each of the key players was responsible for different jurisdictions. Hence the Scottish Development Agency was responsible for lowland Scotland, Tayside Regional Council was responsible for Tayside region (which comprised three district authorities), and only Dundee District Council was focused on the city itself. The principal lesson of the Blackness Project, an early area improvement project discussed in Chapter 6), was that co-ordinated and concerted action could create clear benefits for defined localities.

The Dundee Project was initiated in 1982 as a means of addressing the issues of economic development and physical improvement of the city as whole, through a partnership approach. The main aim of the Project was to halt the decline in employment and population in the city. The agents involved comprised the Scottish Development Agency, Tayside Regional Council and Dundee District Council, and the Project was operational from 1982 to 1991, though it was initially set up for five years. It was organized around a Steering Committee comprising senior officers of the three partners, local councillors, representatives of local businesses and trade unions. It was administered by the Scottish Development Agency, and staffed by Scottish Development Agency staff and secondees from the two local authorities, though a development officer was also seconded from the private sector. Significantly, the Project involved a clear city-wide approach to regeneration, and it enabled wider access to the resources of the individual partners.

The Dundee Project acknowledged that the principal issue for the city was the vulnerability of its economic base. Essentially, the traditional economic sectors were in decline, and the city was seeking to promote new sectors and activities. These processes of decline and adjustment brought a number of attendant social and environmental problems, and a related issue concerned the negative image of the city which deterred inward investment and inward migration of population. While the Dundee Project was a 'bottom-up' initiative, it also played a role within a broader Scottish programme of economic development initiatives, since it formed one of the Scottish Development Agency's area-based projects involving a multi-dimensional approach with an emphasis on the attraction of high technology industry. These area-based projects resulted from a process whereby the Agency had refined its approach to furthering economic development and employment, industrial efficiency and international competitiveness and environmental improvement from a sector basis to a focus on defined geographical localities.

The Scottish Development Agency's approach emphasised the development of competitive and efficient enterprise, economic diversification and the leveraging of private investment into local development schemes. This followed the SDA's involvement in a major area based scheme in Glasgow (the GEAR Project, which had also involved the Scottish Development Agency in a managerial capacity within a partnership arrangement), and area initiatives became a distinctive policy approach of the Agency. The crucial criterion in the selection of areas was potential for improved performance, and so the SDA did not exclude initiatives in relatively prosperous areas where development opportunities were identified.

Transition to the Dundee Partnership

In 1988, at the end of the original agreement for the Dundee Project, a Project Extension Agreement was signed between the three principal partners in order to enable the Project's work to continue to 1991. Its continued success as a model of partnership in Dundee laid the foundations for the establishment of the Dundee Partnership in 1991. The Dundee Project demonstrated the advantages of the partnership approach, and it brought together three separate organizations with different tasks, responsibilities, priorities and resources, and with their own cycles of accountability and scales of responsibility. The SDA, for example, was focused on Scotland, Tayside Regional Council on the regional economy, and Dundee District Council on the city itself. The partnership approach avoided the possibility of lack of co-ordination between these individual bodies. The Dundee Project focused on specific areas or schemes in the city of Dundee, and this was seen by some as a weakness, since it again reflects the limitations of the area-based approach. At times, there was also an uneasy relationship between the constituent bodies in the Project, resulting from their very specific responsibilities. A drawback in this respect is the fact that each body had to report back to different organizations and different scales.

In 1991, a formal agreement was signed between the three principal partners –
Scottish Enterprise Tayside, Tayside Regional Council and Dundee District Council.
This introduced a different set of partners to the institution. Whilst Tayside Regional
Council and Dundee District Council remained the same, the Scottish Development
Agency was replaced by a new organization, namely Scottish Enterprise. Compared
to the SDA, Scottish Enterprise is a broader, more integrated organization in terms of
its potential to contribute to area regeneration, since it has responsibility for economic
development, enterprise and business development, environmental improvement,
and training.

The Partnership was officer-led, with information passed up and down the
management hierarchy, ensuring that priorities were clearly communicated. It was
initially involved in a number of projects with dedicated groups as follows:

- The Community Regeneration Group (comprising Dundee District Council,
  Tayside Regional Council, SET and Scottish Homes, and led by Dundee
  District Council). This Group encouraged the creation of stable self-sustaining
  community areas, and prepared a strategy document *Towards a community
  regeneration strategy*.
- The Land and Property Group (led by SET and Dundee City Council), which
  sought to further the process of liaison, information exchange, joint strategy
  formulation and co-ordination of joint action projects.
- The City Centre Group (led principally by SET through the environmental
  improvement initiative and Dundee District Council, with private sector
  involvement), which concentrated on environmental improvement in the city
  centre.
- The Tourism Group (led principally by SET and Dundee District Council),
  which sought to develop joint projects on the theme of tourism.
- The Enterprise and Employment Group (led principally by SET), which
  sought to enhance job opportunities for Dundee residents.
- The Marketing and the City of Discovery Campaign Support Group (led
  principally by SET, Tayside Regional Council and Dundee District Council),
  which drew on the City of Discovery Campaign Committee, and sought to
  support the Campaign by means of additional city marketing activity.

These projects were multi-speed, involving leadership by specific partners as
appropriate. Thus individual projects were driven by the most relevant officers in the
most appropriate partner, and this could change over time. The Dundee Partnership
therefore brought together the key players in the city economy, and these key
organizations applied the strategic and local functions of planning, infrastructure
provision, environmental regulation and training and business development. Hence
the Partnership could achieve an integration of the key elements of local economic
development and regeneration. In essence, the involvement of the three key bodies
ensured the management of the key factor markets in the city – land, labour and
capital. Furthermore, the external linkages of the three bodies ensured greater access
to the processes related to inward investment and technology transfer. The other
(and certainly not in any way to be considered inferior) participants in the Dundee

Partnership controlled and managed key areas of the Dundee economy. The Dundee Port Authority, for example, played an earlier active role in the regeneration of the port area of the city; the Dundee Trades Council brought the local trades unions in Dundee into the regeneration process; the Dundee Chamber of Commerce integrated the local business community; Scottish Homes (as was) provided access into national housing priorities; and the MPs/MEPs provide the political context to the work of the partnership.

The roles of Tayside Regional Council and Dundee City Council were particularly important in securing links between local communities and projects. Both bodies were democratically elected organizations and were accountable to the local electorate. This meant that the councillor representatives on the Dundee Partnership were obliged to secure as great a degree of transparency in the operations of the Partnership as possible. The Dundee Partnership therefore involved the representation of local and wider interests at a number of levels. In terms of external linkages from the city, a number of interlocking relationships were established with European groups. The East of Scotland Consortium, for example, comprised local authorities in the eastern seaboard of Scotland discussing matters of mutual interest, and the Dundee Partnership was represented through the participation of representatives of the local authorities.

Partnership post-1996

In 1996 there was a reorganization of local government in Scotland, which removed the two-tier structure, and replaced Dundee District Council and Tayside Regional Council with the single-tier Dundee City Council. At this time, the Dundee Partnership was restructured to encompass the interests of a broader range of organizations including the Universities (Dundee and Abertay), Dundee College, Scottish Homes (as was) and representatives from the public and private sectors. The work of the Partnership continued without a permanent team. The strongest links were between the principal partners, which have a strategic overview (from their individual perspectives and portfolio of responsibilities) of the changing circumstances of the local city economy, and an understanding and appreciation of the common agenda.

The Dundee Partnership at its inception had a relatively simple organizational structure, which initially involved a Steering Committee, a Senior Officer Group, a Support Group and a number of project action groups. However, following the expansion of the original remit of the Partnership to include an important role in community planning, the Partnership set up three formal bodies, as well as other groups, to take forward the key community planning themes and allow for consideration of cross-cutting issues. The three formal bodies comprise the Dundee Partnership Forum, the Dundee Partnership Management Group, and the Dundee Partnership Co-ordinating Group. The Dundee Partnership Forum has a broad membership and is chaired by the Leader of Dundee City Council. It meets twice a year, and these meetings comprise participatory workshops on key strategic issues such as transportation and health inequalities.

The Dundee Partnership Management Group consists of the Chief Executives and key officers of the public sector organizations, together with the chairs of each theme group as well as representatives from the private, community and voluntary sectors. The aim of this group is to develop the overall strategy and key priorities of the Partnership, and maximise inter-agency co-operation. The Group meets four times a year, and it is chaired by the Chief Executive of the City Council.

The Dundee Partnership Co-ordinating Group aims to co-ordinate the implementation of community planning, and it includes a representative of each public sector partner and the chairs of the theme groups. It is chaired by an Assistant Chief Executive of the City Council. This structure allows for broader representation than the original structure, but also ensures the effective integration of the Partnership's activities. In addition, there are strategic theme groups covering the themes of 'building stronger communities', 'community safety', 'Dundee's environment', 'health and care', and learning and working'. There are also cross-cutting issue groups covering equalities, community involvement, communication/ marketing, social inclusion/anti-poverty, research and information, and the digital e-city.

Significantly, in 2004, the Dundee Partnership agreed a vision statement to guide its operations and activities. This statement sets out the contextual framework to the work of the Partnership, and it establishes a common agenda to the work of the individual members and partners. The *Vision* expresses this in terms of the constituent key aims and tasks, and it sets out the current actions specific to individual aims, identifies the gaps involved, and sets out the priorities for action to be carried forward by the individual partners. It asserts that the Dundee Partnership strives to achieve a thriving regional shopping, service and employment centre for those living and working in the city, as well as for visitors. The *Vision* therefore reflects a range of overlapping concerns. Hence, for example, the issue of image does not stand alone as a specific item of policy concern and action, since it runs through the objectives or tasks of the Dundee Partnership as a central theme. Indeed, image remains a fundamental issue for many of the activities to which the Partnership addresses itself.

The role of the Dundee Partnership has subsequently evolved and broadened further to become the vehicle for the delivery of Dundee's first Community Plan, and the need to ensure that community planning in Dundee is well managed and implemented now shapes the way the Partnership operates. An important recognition of the importance of the Partnership was reflected in a partnering agreement which was set out in the city's *2005 Community Plan* (Dundee Partnership 2005). This agreement demonstrates the commitment to work together to provide quality services; combine resources to maximise benefits; work together on community consultation; share information wherever possible; promote the values of social inclusion, active citizenship, lifelong learning and sustainability; work to help communities create their own solution to problems; implement action in the community plan; monitor the progress of the Community Plan; and involve the community in the evaluation of the Community Plan.

While the Dundee Partnership is broad and inclusive in operation, there are also parallel linkages with local stakeholders and communities. In terms of linkages within the city, there are a number of avenues open to communities and residents in Dundee. First, there is a local political element via the councillors, the universities and the political representatives. Second, voluntary and community representation is secured through individual projects, for instance in the community regeneration of specific neighbourhoods in the city, or the involvement of small tradesmen in the city centre. Third, local consultation and communication mechanisms are available. In the early stages of the Dundee Partnership, for example, newsletters were prepared and circulated in Dundee, though these were eventually stopped because the individual organizations produced their own documentation. Dundee District Council (as was) in particular had its own newsletter disseminating progress on various initiatives and schemes throughout the city. It was also felt that the publication of a newsletter by the Partnership might deflect attention away from the Dundee economy and onto the Partnership itself. Self-publication in this way was felt to be counter-productive, and the Partnership chose to rely on its established lines of communication.

Evaluation

The Dundee Partnership emerged as a response to the structural and locational characteristics and conditions of the city, and it evolved as a consequence of a number of changes. It provided an important mechanism for establishing a common agenda for the recovery of Dundee under which the different bodies engaged in the process were aware of the priorities being adopted. It brought together the key players, and provided a foundation of understanding to co-ordinate activities, investment and action in defined areas, sectors or issues. It is important to acknowledge that in the absence of the Partnership the organizations involved would have acted much more in isolation, and lines of communication between them would have been less clearly defined level, so arguably there would be greater scope for overlap, policy confusion and the poor phasing and timing of individual programmes.

In spite of the increased complexity of structure that the Partnership has adopted, allowing for consideration of issues between layers and sectors, the Dundee Partnership continues to present an informal institutional arrangement which holds the principal players in a coalition working to an agreed agenda of common priorities. Hence individual actions are taken in the full knowledge of other activities. This suggests that the value added is positive in a number of ways:

- by establishing a more consistent and focused emphasis on the problems of the Dundee economy;
- by creating a momentum of action to address the problems of Dundee as a whole;
- by addressing the image problems of the city;
- by setting a consensus in common priorities for action;
- by avoiding conflict arising from the individual bodies operating on their own, responding to their own defined responsibilities and budgets and operating within their own organizational cultures; and
- by retaining the future needs of the Dundee economy and society at the forefront of thinking by all those involved in the governance of the city – hence the Partnership can create an environment in which attempts to secure institutional territoriality can be resisted by reference to a common agenda.

While the Partnership is public sector-led, there are strong links to the business community both formally through SET (itself a private company) and informally through the work of the local authorities. Given Dundee's economic history, and the severity of its industrial and corporate restructuring, the role of the public sector is very important in asserting leadership for local recovery (Fernie and McCarthy 2001). Moreover, well established lines of communication between individual officers serving as representatives on the Partnership have been developed. However, perhaps the culture of a common agenda is the most important outcome from the work of the Partnership, and there is a clear understanding that the corporate outcome of its work is the overall recovery of Dundee for the benefit of all its residents.

In terms of specific outcomes, in addition to the initiatives outlined above, the Partnership has taken a leading role in joint initiatives aimed at encouraging sectoral business growth in the city, including the following: BioDundee (a collaboration between the public, private and academic sectors to promote the Life Sciences sector), Talking Tayside (a public/private sector initiative to promote the contact centre industry), Tayscreen (a joint initiative to promote film and media production), Business Gateway International (Business Gateway International Trade Tayside offers assistance for identifying contacts and research in target markets), and Interactive Tayside (a public, private and academic partnership to stimulate and diversity the burgeoning digital media industry in the city).

Overall, therefore, the Dundee Partnership continues to represent a means of securing the informal co-ordination of the priorities and activities of its constituent participants, and it has evolved in order to meet changing needs and a changing institutional context (Fernie and McCarthy 2001). While it is not a legal entity, and does not have executive powers or dedicated resources of its own, it seeks, by persuasion, to establish a common agenda for action and spending of the individual participants. Hence it has access to the resources of the constituent members for local authority activities including environmental improvement, economic development, education and social work. It can also participate through Scottish Enterprise Tayside in the wider Scottish Enterprise network. In other words, the Dundee Partnership acts as an important conduit of priorities for policy formulation and implementation towards the common objective of the regeneration of Dundee.

The Social Inclusion Partnerships (SIPs)

As indicated in Chapter 4, Social Inclusion Partnerships (SIPs) reflected the Scottish Executive's emphasis on social justice (Scottish Executive 2002c), and they were developed as an attempt to focus urban regeneration policy on the causes of urban decline rather than its symptoms. As such, the initiative appears to have represented a significant advance on what was previously applied. The designation of SIPs was based upon the use of key indicators that were seen to characterize communities as disadvantaged. These indicators included youth and long-term unemployment, the incidence of low income households, the level of uptake of benefits and education support grants, levels of educational attainment, crime, mortality rates, and other health indicators. Within the SIP areas designated in Dundee, all these indicators were significantly worse than in the city as a whole.

The SIP areas therefore represented the areas of the highest priority in the city for regeneration action, as indicated by the levels of concentration of disadvantage. In addition, however, action in these areas was set within a strategic, city-wide framework for intervention. They also focused upon the most disadvantaged people in society, as well as on the co-ordination of existing programmes and the filling of relevant gaps in provision. As such, they attempted to address directly the issue of social exclusion. This was seen to be a critical element, and a key aspect of the work of the SIPs was seen to be the application of measures that could 'cut across' defined areas of need, and address the broad processes of social exclusion within a city-wide context. In these respects, the SIP initiative can be seen to present an enhancement of previous urban policy initiatives in Scotland.

The SIP initiative in Dundee comprised three elements: SIP1, SIP2 (both 'area-based' SIPs), and 'thematic' SIPs. While the SIP1 areas were defined and coherent areas where disadvantage was concentrated, the SIP2 areas represented those enumeration districts within the worst 10% in Scotland that fell outside the SIP1 areas; they were therefore much more fragmented and dispersed than SIP1 areas, and suffered from a lack of a clear identity. The 'thematic' SIPs were client-based programmes that operated at a city-wide scale, targeting groups such as young people, rather than areas.

The work of the Dundee Partnership, as set out previously, was crucial to the operation of the SIPs in Dundee. More specifically, the Community Regeneration Group within the Dundee Partnership was responsible for social, economic and physical regeneration in the city, and in this respect it facilitated the way in which the constituent partners could work in an integrated fashion so as to achieve broad regeneration outcomes for the city as a whole. However, a range of other actors and groups were involved in the work of the SIPs in Dundee. These included the SIP Team, comprising a SIP co-ordinator and four SIP workers, which had a geographical remit that covered the four SIP1 neighbourhoods, namely Ardler, Kirkton, Mid Craigie and Linlathen, and Hilltown. This team provided the support mechanism for the SIP areas in relation to community empowerment. In addition, four Neighbourhood Development Groups (one representing each of the SIP1 neighbourhoods) formed part of arrangements for developing and implementing relevant area strategies and action plans. The Community and Voluntary Sector Association (CAVA) represented voluntary sector agencies, and there was also a voluntary sector lead officer who provided a link between the voluntary sector and the Community Regeneration Group. In addition, the Community Co-ordinating Group comprised four community representatives for each of the SIP1 neighbourhoods, and this Group was represented on the Dundee Partnership's Community Regeneration Group.

The SIP1 areas incorporated consideration of a number of separate themes and measures. These formed part of the Dundee Partnership Community Regeneration Group's strategic vision for regeneration, which essentially aimed to ensure the creation of stable, sustainable and empowered communities within the city. This vision contained four themes, namely stability, sustainability, empowerment and prosperity. The theme of stability related to measures that sought to, for instance, increase the housing tenure mix, create additional opportunities for private sector involvement, and promote security for businesses and homeowners. The theme of sustainability involved the need to create a sense of belonging for instance by the development of cultural activities as well as educational and training initiatives, and to promote community integration by support for community-led projects. The theme of empowerment involved the need to build community capacity by means of training and skills development as well as access to services and increased citizen involvement by a range of methods. Finally, the theme of prosperity involved measures to enhance the employment prospects of residents in the city by for instance the development of employment and training measures as well as part-time study options.

In fact, the SIP approach may be seen to have emerged directly from the previous Priority Partnership Areas (PPA) approach. Essentially, the Dundee Partnership in 1996 submitted three bids under the PPA programme. These comprised PPA1 (an area of 24,000 people, incorporating 31% of the most disadvantaged enumeration districts in Dundee); PPA2 (an area of 11,200 people, incorporating 18% of the most disadvantaged enumeration districts); and a city-wide Regeneration Programme targeted at 35,000 people, covering 51% of the most disadvantaged enumeration districts in the city. These bids were prioritised, with the PPA1 bid being the top priority, followed by the city-wide Regeneration Programme and then PPA2. The latter was in fact unsuccessful, and some aspects of that bid were incorporated into the successful city-wide Regeneration Programme element. The PPA initiative in Dundee, for example, was based on an archipelago of distinct locations and consisted of four community areas: Hilltown - Maxwelltown, Kirkton, Midcraigie and Linlathen and Ardler. Taken together, these areas contained 31% of Dundee's enumeration districts which fell within the worst 10% of deprived enumeration districts in Scotland. The population within the PPA area was in steady decline in the mid-1990s, and stood at about 15,000 in 2004.

However, in 1998, following a change in government, the Programme for Partnership initiative evolved into the Social Inclusion Partnership initiative, which was held to be more consistent with the new social justice agenda. Broadly, this indicated a shift in approach from a primarily-physical approach to regeneration to one that was more focused on social and economic elements. Nevertheless, the SIP approach, like that of the PPAs, was still broadly area-based in character and orientation.

The SIP1 programme area comprised an archipelago of four very different areas, and so it incorporated clusters of exclusion spread over a large area. The 1991 Census suggested that the nature of disadvantage in this area was multi-faceted and deep-rooted, and the four areas were consistently identified as neighbourhoods in need of an integrated approach to integration, within the Dundee Partnership's Community Regeneration Strategy. The 1991 Census showed that the pattern of the social, economic and physical conditions in the SIP1 area reflected a worsening of conditions, with each of the indicators of exclusion and disadvantage being above the national average. In particular, indicators of unemployment, low-income households, elderly households, and overcrowding, were especially high. In addition, levels of car ownership were low. The baseline year that was set for the SIP1 areas (1996) showed that the areas exhibited widespread symptoms of poverty and exclusion, which were compounded by the physical conditions. For instance, levels of the permanently sick were above the city-wide and national levels. In addition, the level of voids was above the average for the city as a whole, since a substantial surplus in local authority housing (particularly in relation to flats and multi-storey developments) in the city as a whole had arisen from persistent population loss from the city (Fernie 2000).

The population of the SIP areas, based on the 1991 Census, was 14,727. This figure relates to those living in the areas designated, which was shaped so as to ensure that 60% of their enumeration districts fell into the 'most deprived' enumeration districts in Scotland, to aid the process of bidding for funds. The wider neighbourhood, however, was significantly more extensive, covering around 25,000 people, roughly approximating to one-sixth of the population of the city.

The initial funding submission for the PPA/SIP was required to address the broad nature of the problems of the area. In particular, it had to address the fact that 31% of the most disadvantaged enumeration districts in Dundee fell within the area, which also had a significant proportion of the unemployed people in the city. The SIP areas also suffered from broader trends in relation to population loss and instability, with a high degree of movement of households within and away from these areas. Moreover, the Hilltown area contained one of the largest concentrations of ethnic minority residents in the city. In addition, levels of public sector housing in the SIP area were well above the national average, as were levels of low income and one parent families. There were also high levels of uptake of free school meals and school clothing grants, and educational attainment was at a low level. Furthermore, in terms of retail services, there was evidence of a poor retail environment, compounded by high vacancy rates for retail property, and high rates of vandalism. Mortality rates and hospital discharge ratios were also higher than the national average, and there were high and rising levels of reported crime and fire service call-outs.

In terms of the four constituent areas, the first three (Mid Craigie and Linlathen; Kirkton; and Ardler) were relatively homogeneous and predominantly dormitory in character, in that they exhibited the typical characteristics of housing schemes, with relatively few industrial or commercial premises. At the outset of the SIP period, their physical appearance was stark, particularly in those pockets which showed evidence of profound degradation, transient population and unstable communities (Fernie 2000). The fourth area, Hilltown, was different in that it was located closer to the city centre; it also had a broader mix of housing, shops and small factory units. The nature of the areas is set out in more detail below.

Mid Craigie and Linlathen

This area, situated in the Dundee North East area, was largely composed of 1930s or 1940s housing developments, though it contained a mixture of housing styles, and catered for small, medium and large families which required a choice of accommodation including flatted, terraced and semi-detached housing. In fact there had been extensive demolition of housing in the area in the five-year period preceding the establishment of the SIP. Partly as a consequence of SIP designation, the area benefited from physical regeneration including the building of 250 new build low-cost units.

Kirkton

This area, situated in the Dundee North West area, was largely composed of developments built in the 1940s and 1950s. It provided a mixture of housing styles, catering for medium and large families, with housing consisting of flatted, deck access, multi-storey, terraced and semi-detached accommodation. The scope for large-scale physical transformation of the area was not considered to be as significant as in the case of Mid Craigie and Linlathen.

Ardler

This area, in Dundee West, was made up of housing built in the 1960s and 1970s. It provided a mixture of housing styles, catering for small and medium-sized families, with a choice of housing types comprising flatted, multi-story and terraced housing. Much demolition work had taken place in the area prior to designation of the SIP, and it was anticipated that SIP designation would lead to further demolition of multi-story flats that were unpopular and in low demand. In overall terms, the area was considered to be at an early stage of physical redevelopment and regeneration.

Hilltown

This area, located in Dundee Central, was somewhat different to the others in that it was closer to the city centre and contained a mixture of housing, retail, commercial business and factory uses. Predominantly, however, it consisted of housing that had its origins in a much broader period that in the other three cases, spanning a period from the eighteenth century. It provided a mixture of housing styles, catering for individuals and small to medium-sized families, with a choice of housing comprising flatted, deck access, multi-storey, semi-detached and terraced housing. Although this area exhibited a similar character of deprivation and disadvantage as the others in the SIP group, it also displayed clear characteristics of inner city dereliction and exclusion. Partly as a consequence of SIP designation, significant public sector investment was subsequently completed in the area.

Clearly, therefore, in many respects the SIP areas outlined above were very different, and they were subject to considerable change in the first five years of the SIP programme. While the change that took place in these areas is very different in each case, in overall terms it seems clear that the SIP1 programme established appropriate working arrangements for community regeneration, and provided a means of achieving the implementation of a coherent city-wide regeneration strategy in line with the key themes identified by the Dundee Partnership. Thus the SIP1 programmes were integrated with mainstream service programmes, and they also involved substantial engagement with local communities. Again, the Dundee Partnership would seem to have been critical in providing the means by which the aspirations for regeneration could be achieved.

In terms of the SIP2 initiative, the end-term evaluation for this suggested that the dominance of Whitfield within these areas, together with a lack of formal structures, had been problematic. Overall, there was a lack of community identity which was in part a result of the fragmentation of the area. In fact, the end term evaluation suggested that the main characteristics of social exclusion had only been partially addressed, and that such exclusion had been exacerbated by wider population movement across the city as a whole.

Overall, however, the SIP initiatives in Dundee proved a critical means of tackling problems of poverty and disadvantage, and they promoted the involvement and empowerment of relevant local communities, which had a clear and direct influence over decisions in relation to all aspects of funding, prioritisation and service delivery in relation to the SIPs. Indeed, Dundee's provision for community involvement in the SIP process went further than most other SIPs in Scotland, for instance in giving community representatives a majority on many decision-making structures. This necessitated a relatively 'light touch' on the part of the accountable body, Dundee City Council, particularly in relation to the 'geographic' SIPs which contained cohesive communities with a strong sense of identity and active community representatives. In the case of the 'city-wide' thematic SIPs, effective community representation was more problematic because there was not the same clear linkage with specific local communities. Nevertheless, the thematic SIPs were successful in engaging with so-called 'difficult to reach' groups such as young people and young carers, who were less able to participate in formal partnership structures. As indicated below, the work of the SIPs in Dundee was integrated into the community planning process, and dedicated SIP funding was replaced by the Community Regeneration Fund.

The Community Regeneration Fund

As indicated in Chapter 4, the Community Regeneration Fund (CRF) was established by the Scottish Executive in 2004 to bring improvements to the most disadvantaged areas in Scotland. Thus the CRF replaced the SIP Fund, as well as the Better Neighbourhood Services Fund and the Tackling Drugs Misuse Fund. The CRF combined and replaced previous funding mechanisms with similar aims, in particular the Social Inclusion Partnerships, and, like such mechanisms, it targets people in most need. In particular, it is targeted to those communities highlighted as suffering from disadvantage as indicated in the Scottish Index of Multiple Deprivation, and two-thirds of the Community Regeneration Fund was initially targeted on those data zones (areas of around 750 people which the Scottish Index of Multiple Deprivation ranks according to level of deprivation) that represented the most deprived 15% in Scotland. The remaining part of the Fund in 2004 was allocated to those Community Planning Partnerships (see below) with an above average (interpreted as more than 15%) concentration of deprivation in their area. Transitional procedures were put in place to allow smooth changeover to the CRF, and, in 2005/06, Dundee City was allocated £5,775 million from the overall Community Regeneration Fund of £104 million.

The *City Vision*

As indicated in Chapter 4, the 'Cities Review', in Scotland in 2002, aimed to consider economic, environmental and social issues in the cities concerned. In terms of Dundee, the Cities Review Scottish Executive policy response reflects the progress that had been made in terms of economic performance and external image, and it acknowledges the strategic investments in the cultural, retail and public realms. Nevertheless, it also highlights the fact that many manufacturing industries had experienced problems in making a transition to new technologies and markets, leaving a legacy of unemployment and social deprivation as well as environmental problems in relation for instance to derelict land. Significantly, it indicates that, in facing the twin problems of declining population and household numbers, the city was unique in Scotland. It therefore sets out a range of key priorities. Short-term priorities include building on the successes achieved in the biotechnology and games software sectors, enhancing the city's image, and enhancing skills and capabilities. In the long term, priorities include linking to the wider central belt economy. The policy response for Dundee also includes an allocation of £9.3 million from the City Growth Fund, established for all the cities considered. In addition, it suggests the acceleration of land renewal, to be assisted by the allocation of an extra £4 million over three years (Scottish Executive 2003c).

Significantly, allocations under the City Growth Fund were dependent on the creation of strategic city-region agreements, as set out in a 'city vision', in order to set clear priorities for the use of the City Growth Fund. These city visions were to be ready by 2003. Consequently, a ten-year city vision for Dundee was established by the work of the Local Economic Forum and the Dundee Partnership, via the community planning process outlined below. This new *City Vision* updated the city's aspirations and built upon its previous achievements. It was based in part also upon a workshop organised by the Dundee Partnership, which considered the future of the city and the issues that it faced. In addition, meetings were held with representatives of the neighbouring community planning partnerships in Angus, Perth and Kinross, and Fife. While this led to a general agreement on the principles of the *City Vision*, neighbouring Community Planning Partnerships were not prepared to support the view of the Dundee Partnership with respect to local authority boundaries, and so it was agreed that this issue would be addressed separately in an Appendix to the *City Vision* produced by the Dundee Partnership.

In overall terms, the *City Vision* reflects previous aspirations in terms of providing a vibrant and attractive city with a high quality of life, addressing the root causes of social and economic exclusion, and providing a strong and sustainable city economy. It highlights objectives for the city-region in terms of addressing the need for more economic opportunities, and addressing the issue of population loss. The *City Vision* aspires to enhance Dundee's role as a regional centre, enhance the identity of the close network of towns in the surrounding travel-to-work area, and support development that sustained viable communities. Significantly, therefore, the *City Vision* is set within a clear and coherent city-region context, reflecting the new emphasis of the Scottish Executive upon city-regions. This emphasis was reflected not just in the Cities Review document, but also in the wider spatial planning agenda as set out by the Executive (Lloyd 2003).

The key visionary themes established in the *City Vision* are: an enterprising city; a learning city; an inclusive city; a sustainable city; a healthy and caring city; a safe city; and a whole city. The latter aspiration is interesting since it reflects the concern in the city over the governance arrangements allowed by such a tight urban boundary. This is also reflected in the Appendix to the *City Vision*, which shows how the city's local authority boundaries shrunk significantly as a consequence of local government reorganization in 1996. This, it is contended, severely constrains the city's development, administration and public funding. It goes on to suggest that a causal relationship can be seen between constrained local authority boundaries, the under-funding of local government and the social exclusion that is experienced by a high proportion of the city's population. Indeed, it suggests that the city's Council Tax would be significantly lower if the original boundaries had been retained.

A series of outcomes was included for each theme, in terms of the core areas of concern of the *City Vision*, to provide the basis for monitoring. These include halting population decline and achieving a stable and balanced population by 2006; increasing the number of jobs in the city by 1% each year over the following five years; and increasing the average gross weekly earnings in Dundee up to the Scottish average by 2006 (Dundee City Council 2003a).

Community planning

The introduction of community planning in Scotland is outlined in Chapter 4, which also sets out the role of Community Planning Partnerships in bringing together a range of public sector providers such as local authorities, the National Health Service, police, fire services and the local enterprise networks, as well as the communities served by such bodies. The work of the Social Inclusion Partnerships has also been integrated into the community planning process.

The *Community Plan for Dundee 2005-10* (Dundee Partnership 2005) reflects the vision for Dundee set out in the *City Vision* (Dundee City Council 2003a). It therefore reiterates the need to ensure that Dundee is a vibrant and attractive city which offers real choice and opportunity, and which has a strong and sustainable economy, but which also tackles the root causes of social and economic exclusion. It indicates the three objectives that underpin all of the Dundee Partnership's activity to be social inclusion, sustainability and active citizenship. The crucial importance of social inclusion is therefore clear, and this follows the approach of the first community plan, produced in 2001.

After the first community plan, the strategic themes in connection with community planning in Dundee were revised to encompass: building stronger communities (in terms of the regeneration of communities as well as the creation of attractive and popular neighbourhoods); community safety (crime reduction); environment (environmental improvement and protection); health and care (promotion of physical and mental health); lifelong learning (promotion of learning opportunities); and work and enterprise (creation of a vibrant economy).

As in the case of the first community plan, the Dundee Partnership played a key role in the delivery of the plan's objectives. The Partnership created a culture of collaboration, but in a context where there was clear evidence of increasing innovation and wealth, but also deprivation and disadvantage. By 2005, the Dundee Partnership was able to provide clear direction, based on thematic strategies and cross-cutting issues, and it was able to address issues affecting the whole of the Dundee city-region by collaborating with adjoining community planning partnerships. It was helpful in this context that the formal management structure of the Partnership had evolved significantly so as to allow more effective management and an increased number of stakeholders.

The Community Plan acknowledges the importance of the physical regeneration of the city, particularly the regeneration of the waterfront, where a £270 million project is to be carried out over a 15 year period, as indicated previously. The Plan also highlights the importance of community regeneration, and it sets out a number of key objectives in relation to the six areas of activity. First, in relation to building strong, safe communities, the Plan aims to increase the quality and variety of affordable homes, and increase resident satisfaction with the quality and access of local services. Second, in relation with getting people to work, the Plan aims to increase the number of economically active people, and the proportion of 16-19 year olds in training, education or employment. Third, in connection with improving health, the Plan aims to improve the sexual health of young people and reduce levels of smoking, alcohol and substance misuse. Fourth, in connection with raising education achievement, the Plan aims to increase the attainment of qualifications and skills, and the proportion of school leavers entering further or higher education. Fifth, in connection with engaging young people, the Plan aims to increase the availability of and participation in activities for children and young people, and the influence of children and young people in decision-making. Sixth, in relation to effective community engagement, the Plan aims to increase the engagement with minority, vulnerable or excluded groups, and the level of participation on community and voluntary activity.

Such aims are to be achieved in part by the use of the £17.4 million share of the Community Regeneration Fund that was allocated to Dundee, to be used over a three-year period to support those neighbourhoods in most need, building on the progress made by the Social Inclusion Partnership and the Neighbourhood Services Fund. Significantly, the Dundee Partnership also indicated a commitment to the use of Local Community Regeneration Forums that would fund projects which most directly met their own needs in each of the priority areas. These areas comprised: Menzieshill/Charleston/Lochee/Beechwood; St Mary's/Ardler/Kirkton; Hilltown/Stobswell/Fairmuir; Mill O'Mains/Fintry/Whitfield; and Mid Craigie/Linlathen/Douglas.

The Community Plan indicates that the problem of declining population remains one of the key issues facing the city. The city's population was estimated in 2005 at 143,090, but this has been in decline since the 1970s. In particular, since 1993, there have been consistently more deaths than births, though migration has been a more critical determinant of the population, and the assumptions in 2005 were for a net outward migration of around 1,000 people per annum. Moreover, it is significant that the city lost approximately 75% of its land area as well as 17,000 people as a consequence of local government reorganization in 1996. Consequently, in 2005 the city had the tightest administrative boundary of any local authority in Scotland. What are perhaps of even more significance are the changes that have been experienced in the structure of the population, since these have particular significance for service provision. Essentially, from 1991-2005 there was a decrease in 8% in the number of children under 15 and an increase of 11% in the number of people aged 75 or over. Furthermore, Dundee has a lower proportion of 24-35 year olds than any other Scottish city. While there is a high proportion of 18-20 year olds, corresponding with the student population, this falls off sharply since many of this group leave to seek employment elsewhere (Dundee Partnership 2005).

It is also significant is that many people choose to leave the city, but also to say within commuting distance, since the city has a 'daytime' population that is over 10% higher than the resident population. This is the result of around 18,000 people who commute into Dundee to work or study, indicating that city remains a vital regional centre, providing employment and services to a catchment population of around 400,000 (Dundee Partnership 2005). As to projections for the future, it is predicted by the Registrar General that the city's population will be 123,506 in 2018, with younger people accounting for most of this loss, so that it is predicted that nearly half of the city's population by 2018 will be aged 45 or over. Crucially, these projections assume a decrease of 10% in the percentage of the population that is economically active, since this has implications for a whole range of services. In addition, a decline in the number of households is expected.

The city's economy is also an area of concern for the Community Plan, though it indicates that the regeneration of the city had continued since the first community plan. In particular, there was a substantial period of job growth between 2001 and 2004, with a gain of 2,522 jobs. As a consequence, in 2005 the manufacturing sector was only the fourth largest employment sector in the city. The key sectors of growth were retail and wholesaling, leisure and visitor services, public administration, education, health and social work. In addition, however, the city had a strengthening reputation in life sciences and digital media. Moreover, the number of unemployed people fell by 27% between 2001 and 2004, with a 55% fall in the number of long-term unemployed. Nevertheless, the Plan indicates little scope for complacency since the city's overall rate of unemployment was 3.8% in 2005, which was higher than the Scottish average of 2.7%, as well as the UK average of 2.2%. Furthermore, Dundee's unemployment rate was the sixth highest of all of the 32 Scottish local authorities. Nevertheless, gross weekly earnings increased by 25% from 2001-04, and median earnings improved from being 3.8% below the Scottish average in 2001 to 3.9% above the Scottish average in 2005, indicating that the quality of jobs had improved (Dundee Partnership 2005).

In terms of education and skills, attainment in the city's schools was below the national average in 2005, and Dundee also at this time had a higher percentage than the national average of school leavers who become unemployed. In terms of inequality, it is significant that over 25% of the city's population lived in the 15% most deprived data zones in Scotland, and this proportion was only exceeded by Glasgow and Inverclyde. Dundee also had a higher proportion of semi-skilled and unskilled manual workers than the Scottish average. Moreover, only 53.6% of people in the city owned their own home in 2005, as compared to a corresponding figure of 62.6% for Scotland as a whole. In terms of health, life expectancy in Dundee, at around 72 years for males and 78 for females, was below the Scottish average of 73 years for males and 79 years for females. The rate of teenage pregnancies was also the highest in Scotland. In addition, there was a high incidence of smoking in the city, with around two-fifths of the city's population being smokers, and there was a high correlation between levels of smoking and areas of concentrated disadvantage (Dundee Partnership 2005).

The Plan also indicates the work done by the Dundee Partnership in engaging with communities. It shows that the Community Involvement Strategy, agreed and revised in 2005, had led to progress in consolidating partnership activity by the stakeholders involved in the Dundee Partnership. It also indicates that the Social Inclusion Partnership Boards played a crucial role in the achievements of the SIP projects, and that the Dundee Partnership was successful in introducing a flexible model of community representation that extended the rights and responsibilities of Community Councils, to other less formal organizations. In addition, Local Community Plans were developed for each of the nine localities in Dundee. These Plans had a timeframe of three years, and they combined what had been the responsibility of plans and initiatives such as Community Learning Plans, the Better Neighbourhood Services Fund and the Area Regeneration Plans for the geographic Social Inclusion Partnerships. Local Community Regeneration Forums were also established, as indicated previously, to empower communities in the Community Regeneration Areas. The intention was that these would allow local people to determine exactly how Community Regeneration Fund projects would be allocated (Dundee Partnership 2005).

In terms of building stronger communities, the Community Plan shows that a clearer connection had been made between community planning and the neighbourhoods. It also indicates that significant progress had been made in enhancing community safety, though further progress was needed in terms of tackling substance and alcohol misuse for instance. In terms of the environment, again key achievements had been made since the last community plan for instance in terms of restoration of Baxter Park, and new themes and priorities had been developed via the Dundee Partnership in terms of education and environmental responsibility, improving local environments (particularly in community regeneration areas), minimising pollution, enhancing sustainable transport, protecting biodiversity, using energy wisely, and managing waste.

In terms of health and care, the Plan shows that the *Joint Health Improvement Plan 2005-08*, produced by the Community Planning Health Action Team, had provided a framework for tackling health deficiencies, though further work was needed to tackle severe health inequalities and the implications for particular groups such as ethnic minorities. In terms of lifelong learning, the Plan acknowledges the progress made by the Dundee Partnership, for instance in bringing about a wider range of learning provision by Dundee College and other providers. It also highlights the priorities of community learning and capacity building (particularly in areas suffering from deprivation); services for children; transition from school to adult life (for instance via the Xplore Social Inclusion Partnership; work-based learning (training provision from companies); higher and further education, and learning in later life. In addition, new strategic developments are set out such as promoting self-confidence (particularly for disadvantaged groups); tackling barriers and providing support (for instance by providing networking and personal development opportunities); and pathways and progression (for instance by supporting those facing redundancy).

In terms of work and enterprise, the Plan acknowledges the progress made for instance in bringing about high levels of retailing investment, but it also acknowledges that the new policy emphasis on the city-region, as articulated in Dundee's *City Vision*, implies the need to work towards the strategic goal of enhancing Dundee's role as a strong regional centre. This was seen particularly in terms of providing a major location for employment and investment; encouraging knowledge, innovation and enterprise; and providing a vibrant cultural, leisure and retail centre.

For each of the above themes, the Community Plan provides a series of detailed action points with timescales, with baseline and target indicators, and these results are to be reported back each six months to the Dundee Partnership Management Group. In fact, many of the indicators used in the Plan were included in the Community Regeneration Outcome Agreement, and the monitoring framework for the Partnership set targets for the following three to five years.

Conclusions

Many of the above initiatives, which are largely holistic in approach, involve particularly important social regeneration elements. They show evidence of an evolution and policy learning approach that in several respects mirrors the experience of Scotland as a whole, as set out in Chapter 4. In particular, a focus on social need and social justice, expressed either geographically (as in SIPs 1 and 2) or sectorally (as in the thematic SIPs), seems to run throughout the recent history of regeneration in Dundee in many respects, albeit in the context of the priority of retaining population based on economic growth. It may also be argued that there has been a clearer targeting of funding and resources on such needs than occurred in many contexts in England, and such an emphasis may derive in part from the cross-cutting policy approach of the Scottish Executive on the promotion of social inclusion. Evidence of both policy and institutional learning can also be seen for instance in terms of the integration of policy in community planning.

Moreover, the reflective approach of the Dundee Partnership, illustrating the capacity to evolve based on experience, may provide in many respects a model for collaborative practice in regeneration. In relation to the concepts set out in Chapter 2, the Dundee Partnership seems to illustrate an urban regime that acknowledges the principles of communicative action, collaborative planning and institutional capacity-building. This may be seen in all aspects of its work, but perhaps most clearly in the community planning function. The success of the Partnership in this respect may derive in part from historical and cultural traditions of partnership and community involvement. For instance, the rich history of trade union activity in the city, led to significant community capacity and enthusiasm for involvement at all levels of policy development and application. In addition, the tradition of municipal leadership meant that the Partnership was strongly public-sector led, enabling a degree of strategy and co-ordination to be consistently applied. Equally, however, the corporate approach to policy development that was a feature of all Scottish cities to some extent, can be seen to have been particularly important in Dundee, with the Dundee Partnership able to include most stakeholders effectively, partly because of the relatively small scale of the city.

However, there remain unresolved questions concerning the practice of broadly-based regeneration, involving economic, environmental and social aspects. This reflects broader tensions between the continued dominance of the need to enhance competitiveness and economic growth on the one hand, and to protect amenity and promote social inclusion on the other. The agenda for modernization of spatial planning in Scotland, for instance, involves an explicit assumption that both aims can be achieved in parallel without contradiction or dilution of either. However, it may also be suggested that this is unrealistic and that one or other of the broad aims must inevitably dominate, with the agenda for competitiveness and growth at present having priority. At the local level, this might be illustrated by the continued emphasis in Dundee on image enhancement as shown for instance by the priority given to physical regeneration of the waterfront area of the city. Nevertheless, these are tensions that perhaps cannot be resolved locally, and in Dundee, the success of strategies for growth of new economic sectors such as the life sciences and digital media, indicate that economic growth and social regeneration can indeed be achieved in parallel.

Again, however, the area-based nature of many aspects of regeneration initiatives in Dundee, particularly the PPA and SIP1 initiatives, bring associated problems of potential displacement rather than resolution of problems, as indicated in Chapter 7. At the same time, however, the geographical targeting and area focus of much regeneration activity in Dundee has been able to build upon local feelings of identity and belonging. Indeed, the dispersed geography of the SIP2 programme led to a frequent criticism was that it was not associated with easily-identifiable communities. Hence it was viewed as 'invisible', with no local community feeling a sense of 'ownership' of the programme. This had a knock-on effect on community involvement, which was low in comparison with many other programmes. Clearly, therefore, the debates on the merits and disadvantages of area-based regeneration initiatives are complex and multi-faceted, though there would seem to be a consensus that such initiatives have a role to play within a broader set of regeneration strategies. There would also now seem to be a consensus on the need for 'holistic' approaches as a necessary condition for sustainable regeneration.

Chapter 9

Conclusions

Introduction

The origins and development of urban regeneration policy in Scotland, and its application to the city of Dundee, has been set out in the preceding chapters. This concluding chapter summarizes urban regeneration achievements and problems for Dundee, presents linkages between the experience of Dundee and the conceptual frameworks set out earlier, and highlights the main unresolved issues and tensions for urban regeneration in Scotland and beyond.

Achievements and problems

It is clear that a number of regeneration achievements have been made in Dundee in recent decades, in parallel with similar achievements in other cities in Scotland. Previous chapters have summarized the experience of the city in this respect, and achievements in relation to physical, economic and social regeneration have been highlighted. Moreover, in addition to the initiatives described, the city centre has been the subject of major investment in public realm improvements including pedestrianization, street furniture, feature lighting and public art works. As a consequence, the city has received awards from the Royal Town Planning Institute (Award for Planning in Urban Renaissance), the British Urban Regeneration Association (Award for Best Practice in Regeneration), and the British Council of Shopping Centres (Award for Most Improved City Centre). Moreover, such public investment has encouraged major retail development, with the refurbishment of the Wellgate Shopping Centre in 1993 and the opening of the redeveloped Overgate Shopping Centre in 2001.

Furthermore, major tourist attractions have been developed, including Discovery Point (1993), the Sensation Science Centre (2000) and the four-star Apex Hotel (2003). The Dundee Partnership has been critical in bringing about such initiatives, and it has also helped to ensure that the city is now an expanding conference venue and a port of call for cruise liners. The city is also increasingly active in marketing its attractions, with the City of Discovery Campaign using the local, national and international press to ensure that the positive developments taking place are widely communicated. This is seen as essential for the attraction of visitors and inward investment.

Major achievements are also evident in terms of inward investment for economic development, for instance with the development of the £13 million refurbishment of the Wellgate Shopping Centre (1993), the £12 million Wellcome Trust bioCentre

(1997), the £6 million BT Contact Centre (1999), and the £150 million new Overgate Shopping Centre (2000), In addition, the Dundee Partnership, together with other agencies, pursued the Tayside Civil Service Campaign that brought the development of the Scottish Commission for the Regulation of Care and the Scottish Social Services Council (creating 260 jobs) in 2001, the Department of Work and Pensions Pensions Service Centre (creating 550 jobs) in 2002, and the Inland Revenue Contact Centre (creating 470 jobs) also in 2002.

Illustration 9.1 Public art in Dundee: 'Desperate Dan'

Nevertheless, Dundee continues to face major problems in terms of regeneration, as indicated by the 2006 Scottish Index of Multiple Deprivation (Scottish Executive 2006b). This shows that the city continues to exhibit a high degree of income deprivation, with 26,835 people in this category, comprising 18.9% of the population of the Dundee City Council area, compared to the Scottish average of 13.9%. The Dundee figure is the third highest of any local authority in Scotland, after Glasgow (at 24.7%) and Inverclyde (at 19.2%), and the city saw an increase of seven data zones in the 15% most income deprived areas in Scotland between 2004 and 2008. In terms of employment deprivation, 29.6% of the city's data zones fall into the 15% most employment deprived areas in Scotland. Again, this places the Dundee City Council area as the third highest in Scotland, after Glasgow (44.7%) and Inverclyde (41.8%). Moreover, the figure in Dundee had increased from 26.3% in 2004 (though the percentage of working age people who were unemployed had decreased from 17.3% in 2004 to 16.9% in 2006). In terms of health, 24% of data zones in Dundee fall into the 15% most health deprived areas in Scotland, with this again putting the Dundee City Council area third most deprived after Glasgow and Inverclyde. However, Dundee saw a decrease in the percentage of its data zones in the 15% most health deprived in Scotland between 2004 and 2006. In terms of education, training and skills, it saw an increase in the number of data zones in the 15% most education deprived zones in Scotland from 48 in 2004 to 55 in 2006 (Scottish Executive 2006b).

Clearly, therefore, Dundee faces continuing problems in many aspects of social and economic disadvantage, in spite of the major achievements made in terms of regeneration initiatives. Low incomes in particular remain a critical issue, with the situation worsening since 2004 relative to Scotland as a whole. In addition, the ongoing loss of population from the local authority area remains significant, with a loss of over 10% expected between 2002 and 2018 (Scottish Executive 2004b), and it remains a corporate priority of Dundee City Council and the Dundee Partnership to address this issue.

Linkage to conceptual frameworks

It is useful here to summarize the experience of Dundee in terms of regeneration policy and practice with respect to the theoretical and conceptual frameworks introduced in Chapter 2. The experience of Dundee in terms of urban decline and regeneration in some respects accords with that in many other cities in Scotland and elsewhere. While there is clear evidence of post-industrial 'renaissance', there is also a legacy of industrial decline, exacerbated by historic reliance on a narrow industrial base, with contemporary effects in terms of low incomes, educational attainment and other indications of disadvantage and social exclusion. Broader processes of globalization are clearly significant in this respect, though the effects of various national approaches to economic policy have also been felt.

Thus the underlying processes that give rise to problems of urban decline and disadvantage are essentially similar in Dundee to many other cities in the UK and elsewhere. However, what has been different in many respects is the response that has been made to such problems. The institutionalist perspective introduced in Chapter 1 is of clear relevance here, with the interaction of structure and agency assisting in the explanation of processes of decision-making and strategy development in the city. However, it is necessary here also to make reference to the broader national context within which the relatively corporatist approach of governance in Scotland has been evident. Hence the recognition of a need for partnership between public and private sectors, and the application of this in policy terms, has arguably been more evident in Scottish policy and practice than that in many other UK contexts. This would seem to have been in part a consequence of the dense and tightly-knit institutional architecture in relation to governance in Scotland, and the role of the Scottish Parliament seems to have contributed significantly to this. In addition, a relatively interventionist approach has been evident, illustrated for instance by the experience of the Scottish Development Agency. Moreover, initiatives such as the GEAR project may be seen to illustrate a relatively strategic approach to urban regeneration.

These factors of course inform wider areas of concern such as the 'locality debate' referred to in Chapter 2. Again, the national level in Scotland has arguably involved a distinctive approach to the framing of responses to globalizing pressures, as well as a degree of consensus in relation to aspects of regeneration such as a commitment to principles of social justice. This has resulted in a clear strategic policy context as demonstrated for instance by the *National Planning Framework for Scotland* as well as other national policy statements. More recently, policy developments associated with the modernization of the spatial planning agenda, including an emphasis on city-regions, also assist in the framing of a strategic context for regeneration activity, though such strategic approaches may of course be traced back to initiatives such as the Tay Valley Plan.

In addition, locally-distinctive approaches have been evident in Dundee. This is illustrated by the high degree of local corporatism, albeit within a context of local authority leadership that has avoided problems of sectionalism whereby dominant cultures act to prevent change. The approach to decision-making has been inclusive in many respects, with clear consensus-building at the local level. In recent decades, such an approach may be illustrated by the Dundee Partnership. However, again this reflects a complex set of historical and contextual factors that were identified for instance in the *Dobson Chapman Plan* of 1952, which made specific reference to the city's 'corporate consciousness'. More recently, aspects of governance in Dundee may be seen to indicate 'institutional thickness', with a high degree of trust, and shared norms and traditions. Hence, while the approach of entrepreneurial cities and city competition is evident in Dundee as elsewhere, as illustrated by the emphasis on city marketing, this is nevertheless moderated and framed by an acceptance of the need for strong local authority leadership, as well as the clear integration of the Dundee Partnership within local governance decision-making structures. As Carley (2006) notes, unless partnerships are linked to decision-making in this way, their achievements are inherently limited.

Moreover, the Dundee Partnership displays an awareness of the need for institutional capacity, effective partnership networks and clear communicative processes. Hence the Partnership would seem to have enabled a measure of consensus-building that in many other contexts has been much more difficult to achieve, and, in spite of some (relative) shortcomings such as the involvement of the private sector, it seems to have involved and empowered the most relevant stakeholders. Moreover, it shows how networks and structures can be re-negotiated or transformed to adapt to changing circumstances. While the issue of scale in Dundee has been helpful in facilitating what Carley (2006) calls 'partnership rationalisation', whereby one overarching partnership is able to avoid partnership proliferation, cultures of joint working and corporatism have also been critical. Overall, therefore, the distinctive nature of governance processes and structures in Dundee, allowing collaborative action to operate in a context of tightly-knit cultural communities, seems to have been critical in shaping the city's experience of urban regeneration.

Issues and tensions

It is now possible to highlight some unresolved issues and tensions regarding urban regeneration in a broader context, following from conclusions in relation to the experience of urban regeneration in England and Scotland, as well as the particular experience of Dundee. These are set out below.

Need for holistic regeneration approaches

As indicated in Chapters 3 and 4, the experience over recent decades in both England and Scotland has shown that it is not sufficient to address primarily physical problems (though these are often most straightforward). Hence, in the Scottish context, the examples of the Glasgow Eastern Area Renewal (GEAR) project (mentioned in Chapter 4) and the New Life for Urban Scotland (NLUS) initiative (illustrated by the experience of Whitfield in Chapter 7) show the inherent limitations of an approach that does not engage fully with underlying processes of economic and social decline. In England, the Urban Development Corporations (UDCs) would also seem to have failed to ensure economic and social benefits for local communities, though these initiatives did indeed physically transform large areas of vacant, derelict or underused land. As indicated previously, it may be argued that a largely property-led approach to regeneration continues to dominate in some respects in Dundee, as illustrated for instance by the dominance of the central waterfront project in the vision for the city. This project is seen as a major symbol of post-industrial renaissance in the city, but again its physical elements must be combined with social and economic outcomes in order to provide sustainable urban regeneration.

The lesson therefore would seem to be to use physical regeneration as a component – perhaps indeed a central component – of regeneration initiatives, but only in the context of a broader strategy encompassing a 'holistic' approach that also provides social and economic benefits to local communities, particularly those in greatest need. This would seem to be supported by evidence from elsewhere in Europe. For instance, the experience of major urban unrest in Paris in 2005 suggests that the emphasis on physical urban regeneration by the French government, as shown for instance by the Great City Projects, meant that underlying causes of social exclusion, such as poor educational outcomes, were not addressed in many areas. Moreover, the high level of youth unemployment in France, at around 25%, would also seem to have been an important factor in this respect, particularly when compounded by problems of discrimination with regard to young people from ethnic minorities.

Hence regeneration policy and practice should incorporate social and economic aspects and seek to address causes rather than symptoms of decline. A crucial element here is the need to provide appropriate linkages to economic opportunities, which has not always been successfully achieved. For instance, again in the case of the *New Life for Urban Scotland* initiative (illustrated by the case of Whitfield in Chapter 7), in overall terms this did not appear to ensure that local people had access to suitable employment in line with their skills. While the Social Inclusion Partnerships initiative, considered in Chapter 8, clearly attempted to address issues of barriers to work such as childcare and transport, unresolved issues remain in terms of low economic activity rates in cities such as Dundee.

Policy integration

A further lesson from regeneration policy throughout the UK since the 1960s has been the need for more effective integration of economic, social and environmental aspects of regeneration, so that these can be mutually reinforcing. This has been particularly critical in England, as set out in Chapter 3. In Scotland, such problems of integration would seem to have been less evident, in part the result of the reduced scale (in absolute terms) of problems of degeneration, as well as simpler administrative and institutional structures. Nevertheless, as indicated in Chapter 4, there would seem to be scope for improvement in this respect, as well as for the further integration of regeneration policy with other areas of policy.

The Social Inclusion Partnerships (SIPs) initiative did in many respects attempt to ensure an integrated approach at the level of the geographical neighbourhoods involved. However, the Scottish Executive concluded that the SIPs initiative as a whole did not always integrate local regeneration efforts with wider strategic planning and service delivery. While community planning and the production of Regeneration Outcome Agreements (ROAs) is addressing this issue to some extent, it is also necessary to ensure greater integration of local approaches with broader-scale consideration of strategic planning in relation to land use, transport, housing and public services. Here, the role of the *National Planning Framework for Scotland* (Scottish Executive 2004b) is critical, since this links issues of regeneration with broad strategic concerns, for instance for transport and employment, from a national perspective.

It is significant in this context that the English Urban Task Force report suggested the need for a shift in the role of spatial planning, by means of 'a planning system which is strategic, flexible and accountable' (Urban Task Force 1999, 44), with the function of land use planning acting as 'a positive mechanism for achieving change, particularly urban regeneration objectives, rather than, primarily, a reactive means of controlling development' (1999, 44). Clearly, urban regeneration activity must take place within a broader context, and the spatial planning system provides an important means of ensuring greater coherence and integration in many aspects of regeneration.

However, it may also be suggested that in the Scottish context there is a major gap in the institutional infrastructure necessary for policy integration for urban regeneration between scales. This arises from the lack of regional development frameworks or structures to support urban regeneration activity as well as other areas of policy. The contemporary emphasis on city-regions, the possibilities for joint working between authorities, and the increasing importance of national planning, all seek to address this issue, but it may be argued that they do not go far enough.

Partnership

The importance of broadly-based partnership has been a key theme throughout this book. In the arena of regeneration, especially that which is 'holistic' in nature, sustainable outcomes can only be achieved by means of the contribution and commitment of a variety of actors and agencies from the private, public and voluntary sectors. The Social Inclusion Partnerships, for instance, attempted to ensure such a broadly-based partnership approach, and the work of the Dundee Partnership has been critical in harnessing the advantages of partnership to further regeneration activity. In addition, the application of Community Planning Partnerships in Scotland (with the Dundee Partnership playing a critical role in community planning) and Local Strategic Partnerships in England (which play a similar role) indicates that the need for effective partnership is now evident to all.

Nevertheless, a difficult and persistent issue within partnership is that of the effective involvement of local communities. This is now recognised as an important aspiration in order to build in better local knowledge concerning needs, greater potential for the joining up of services, and enhanced motivation of front-line staff that can lead to innovation in service delivery (Scottish Executive 2006). The delivery of this aspiration, however, is often problematic, in view of the limited capacity of communities to contribute, and the persistence of processes that prevent communities exercising real power over decision-making. Despite some difficulties, this would not seem to have been a significant obstacle in Dundee, because of factors such as the strong community infrastructure linked to the city's history of community activity and trade unionism, the continued presence of clearly-defined and cohesive neighbourhoods, and the commitment of the local authority to community involvement.

The approach of the Social Inclusion Partnerships throughout Scotland illustrated a commitment at national level to community involvement as a key element of regeneration, and such a commitment is also reflected in the application of the Regeneration Outcome Agreements (ROAs) as part of the Community Regeneration Fund (CRF). Nevertheless, there remain issues concerning the need for capacity-building to allow local communities to contribute effectively, and the need to ensure that those most active at local level have the broad support of the local population. These issues have been felt to some extent in Dundee, where it has sometimes been difficult to sustain the enthusiasm of local community representatives, particularly given the scale of regeneration activities in the city.

In a wider context, the effective involvement of local communities was highlighted in a report produced by the Scottish Urban Regeneration Forum (SURF) (Scottish Urban Regeneration Forum 2006) which synthesised comments arising from a series of open forums. Participants in these forums indicated the need for real engagement of communities, meaning more than simply selecting from options prepared by others. One implication of such engagement, the report suggests, is the need for extended timescales for participation. While this increases costs, it may be suggested that in the long term it is the most efficient and economic approach to regeneration activity, and is necessary to generate trust as well as to maximise involvement. The alternative approach of aiming at early action may meet mechanistic target-setting agendas, but may also conflict with more strategic and sustainable approaches. In addition, the report argues that it is helpful to focus on processes as well as structures in partnership working. While it is often easier to incorporate structures that are inclusive, the effective implementation of inclusive processes of working is more difficult, but arguably more important. This reflects the discussion of networks in Chapter 2, and the experiences of the Dundee Partnership.

A broader aspect of partnership working arises from the need to develop learning organizations and partnerships that can change culture and implement change. This would seem to be particularly relevant within the Scottish context of community planning, involving new, diverse groupings of agencies in relation to integration of service delivery (Scottish Urban Regeneration Forum 2006). Again, the experience of the Dundee Partnership is salutary in this respect, since it illustrates that institutional learning and evolution is possible. Such learning is necessary also on a wider basis to ensure that the lessons of practice are not overshadowed by the short-term political requirements for new initiatives and priorities.

Effective evaluation

As indicated elsewhere in this book, much has been done to ensure the effective evaluation of initiatives, with a number of reports highlighting the strengths, weaknesses and lessons arising from key regeneration initiatives such as the Whitfield Partnership. However, in view of the continued need to ensure a learning approach, and the persistence of problems such as inadequate policy integration, it seems necessary to give more emphasis within regeneration activity to effective evaluation. In this context, the case may be made for more qualitative approaches that engage with the complexities of context and with the processes of partnership, as opposed to a narrow focus on outcomes, which is often applied in order to comply with funders' requirements (Scottish Urban Regeneration Forum 2006).

Final thoughts

In overall terms, the case for investment in urban regeneration in order to address market failures remains clear. There are linkages here to wider policy concerns such as compact cities and polycentric development, as applied for instance in the broader European Union context, as well as to notions of 'smart growth' as applied in the USA. While such concepts are often presented as only indirectly related to economic performance, it may be asserted that regeneration activity, emphasising compact, mixed-use developments, can foster dense labour markets, vibrant cities and efficient transportation systems, which can all enhance economic performance at regional and national levels. This is linked to the increasing importance of knowledge-based industries and the associated need to improve local amenity so as to attract a mobile and increasingly discriminating workforce. Thus regeneration activity can provide benefits far removed from areas of concentrated disadvantage, and it is therefore of benefit to cities and city-regions as a whole, as well as to regional and national economies.

Moreover, a report by the UK's Office of the Deputy Prime Minister (ODPM) in 2006 concludes that public expenditure on regeneration can bring clear national benefits such as enhancement of the efficiency of the local economy, more effective utilisation of factors of production such as land, and increased national competitiveness and productivity. However, the report also suggests that regeneration activity should be set in the context of a wider economic development strategy, taking into account the functional roles of the cities or areas concerned, and that regeneration policies should be targeted in areas where market failures are most acute (Office of the Deputy Prime Minister 2006). The experience of regeneration in Dundee, as well as more widely in Scotland, would seem to reflect these assertions. In a broader Scottish context, there is clear acknowledgement of the need for a broad strategic context, and this is a central element within much contemporary regeneration and spatial planning policy. In addition, the targeting of initiatives to areas of clear market failure perhaps reflects to a degree the continued emphasis in Scotland on addressing need, albeit in a shifting context within which competitiveness and growth feature increasingly highly, with the need for choice in terms of trade-offs between competing policy priorities often being suppressed. Again, this suggests that regeneration aims must be clear and unambiguous. Only by such means can regeneration policy and practice aspire to address the entrenched problems of urban disadvantage that persist in many Scottish cities and beyond.

Bibliography

Alexander, A. (1997), 'Scotland's Parliament and Scottish local government: conditions for a stable relationship', *Scottish Affairs* 19, 22-28.

Allen, J. (1998), 'Europe of the neighbourhoods: class, citizenship and welfare regimes', in Mandanipour, A., Cars, G. and Allen, J. (eds.).

Amin, A. and Thrift, N. (1995), 'Globalisation, Institutional "Thickness" and the Local Economy', in Healey, P., Cameron, S., Davoudi, S., Graham, S. and Mandanipour, A. (eds.).

Ashworth, G.J. and Voogd, H. (1990), *Selling the City* (London: Belhaven Press).

Atkinson, R. and Moon, G. (1994), *Urban Policy in Britain* (London: Macmillan).

Audit Commission (1989), *Urban Regeneration and Economic Development* (London: Audit Commission).

Bagnasco, A. and Le Gales, P. (eds.) (2000) *Cities in Contemporary Europe* (Cambridge: Cambridge University Press).

Bailey, N., Barker, A. and MacDonald, K. (1995), *Partnership Agencies in British Urban Policy* (London: UCL Press).

Bazley, K. (1993) 'Urban Regeneration and Economic Development', in Duncan of Jordanstone College of Art (1993) *100 Years; town planning in Dundee*, Duncan of Jordanstone College of Art, Dundee.

Bell, D. and Jayne, M. (eds.) (2006), *Small Cities: urban experience beyond the metropolis* (London: Routledge).

Bennett, B. (1995), 'SRB: advice for practitioners', *Planning Week*, 8.6.95, 19.

Berry, J., McGreal, S. and Deddis, B (eds.) (1993) *Urban Regeneration: Property Investment and Development* (London: E and F N Spon).

Bianchini, F. (1993a), 'Remaking European Cities: the role of cultural policies', in Bianchini, F. and Parkinson, M. (eds.).

Bianchini, F. (1993b), 'Culture, conflict and cities: issues and prospects for the 1990s', in Bianchini, F. and Parkinson, M. (eds.).

Bianchini, F. (1996), 'Culture, economic development and the locality: concepts, issues and ideas from European debates', in Hardy, S., Malbou, B. and Taverner, C. (eds.).

Bianchini, F. and Parkinson, M. (eds.) (1993), *Cultural Policy and Urban Regeneration: the West European Experience.* (Manchester: Manchester University Press).

Blackman, T. (1995), *Urban Policy in Practice* (London: Routledge).

Booth, P. and Boyle, R. (1993) 'See Glasgow, See Culture', in Bianchini, F. and Parkinson, M. (eds.).

Boyle, R. (1988), 'Private sector urban regeneration: the Scottish experience', in Brown, A., McCrone, D. and Paterson, L. (eds.).

Boyle, R. (1990), 'Regeneration in Glasgow: Stability, Collaboration and Inequity', in Judd, D. and Parkinson, M. (1990) (eds.).

Brown, A., McCrone, D. and Paterson, L. (1998), *Politics and Society in Scotland* (London: Macmillan).

Brownill, S. (1990), *Developing London's Docklands* (London: Paul Chapman).

Cabinet Office (2006), *Reaching Out: An Action Plan on Social Exclusion* (London: Cabinet Office).

Cambridge Policy Consultants (1999), *An Evaluation of the New Life for Urban Scotland Initiative in Castlemilk, Ferguslie Park, Wester Hailes and Whitfield* (Edinburgh: Scottish Executive).

Cameron, S. and Davoudi, S. (1998) 'Combating social exclusion: looking in or looking out?' in Mandanipour, A., Cars, G. and Allen, J. (eds.).

Campbell, S. and Fainstein, S. (eds.). (1996) *Readings in Planning Theory* (London: Blackwell).

Carley, M. (2000), 'Urban Partnerships, Governance and the Regeneration of Britain's Cities', *International Planning Studies* 5:3, 273-97.

Carley, M. (2006), 'Partnership and statutory local governance in a devolved Scotland', *International Journal of Public Sector Management* 19:3, 250-60.

Clarence, E. and Painter, C. (1998), 'Public services under new Labour; collaborative discourses and local networking', *Public policy and Administration* 13:3, 8-22.

Comedia (1991), *Out of Hours: a study of economic, social and cultural life in 12 towns and cities in the UK* (London: Comedia).

Community Planning Working Group (1998), *Report of the Community Planning Working Group*, July (Edinburgh: Scottish Office).

Corporation of Dundee (1957), *The City and County of Dundee Official Handbook* (Dundee: Corporation of Dundee).

Council for Cultural Co-operation (1995), *Culture and Neighbourhoods Volume 1* (Strasbourg: Council of Europe Publishing).

Cox, K. and Mair, A. (1988), 'Locality and community in the politics of economic development', *Annals of the Association of American Geographers*, 78:2, 307-25.

Cox, K. and Mair, A. (1991), 'From localised social structures to localities as agents', *Environment and Planning A* 23, 197-213.

Cullingworth, J.B. (2002), *Town and Country Planning in Britain* (London: Routledge).

Darlow, A. (1996), 'Cultural Policy and Urban Sustainability: making a missing link?', *Planning Practice and Research* 11, 291-301.

Deakin, N. and Edwards, J. (1993), *The Enterprise Culture and the Inner City* (London: Routledge).

Deas, I. and Harrison, E. (1994), 'Hopes left to crumble', *Local Government Chronicle*, 4.3.94, 13.

Department of the Environment (1977), *Policy for the Inner Cities*, Cmnd 6845 (London: HMSO).

Department of the Environment (1994), *Inner Cities Research Programme: Assessing the Impact of Urban Policy* (London: HMSO).

Department of the Environment, Transport and the Regions (1999), *Area-based and other regeneration-related initiatives: summaries* (London: DETR).

Department of the Environment, Transport and the Regions (2000), *Our Towns and cities: The Future – Delivering an Urban Renaissance* (London: DETR).

Department of Culture, Media and Sport (1999), *Guidance for Local Authorities on Local Cultural Strategies* (London: Department of Culture, Media and Sport).

Dobson Chapman, W. (1952), *The City and Royal Burgh of Dundee: Survey and Plan 1952: Part Two: the city plan* (Macclesfield: Dobson Chapman Partners).

Doherty, J. (1991), 'Dundee. A post-Industrial City', in Whatley, C.A. (ed.).

Donnison, D. (1987), 'Conclusions', in Donnison, D. and Middleton, A.

Donnison, D. and Middleton, A. (1987), *Regenerating the inner city: Glasgow's experience* (London: Routledge and Kegan Paul Ltd.).

Duncan of Jordanstone College of Art (1993), *100 Years; town planning in Dundee* (Dundee: Duncan of Jordanstone College of Art).

Dundee City Council (1996), *Economic Development Plan 1996-99* (Dundee: Dundee City Council).

Dundee City Council (1997), Arts Action Plan 1998-2000 (Dundee: Dundee City Council).

Dundee City Council (2002), *Dundee Central Waterfront Development Masterplan 2001-2031* (Dundee: Dundee City Council).

Dundee City Council (2003a), *Dundee – A City Vision* (Dundee: Dundee City Council).

Dundee City Council (2003b), *Finalised Dundee Local Plan Review* (Dundee: Dundee City Council).

Dundee City Council (2005a), *Cities Growth Fund – Annual Report 2004-5 Dundee Central Waterfront* (Dundee: Dundee City Council).

Dundee City Council (2005b), *Local Housing Strategy Review/Action Plan Update* (Dundee: Dundee City Council).

Dundee City Council (2006), *Local Housing Strategy 2004-2009* (Dundee: Dundee City Council).

Dundee District Council (1979), *Inner City Local Plan* (Dundee: Dundee District Council).

Dundee Partnership (2005), *Community Plan 2005-2010* (Dundee: Dundee City Council).

Ebert, R., Gnad, F. and Kunzmann, K. (1994), *The Creative City* (Stroud: Comedia).

English Partnerships (1994), *English Partnerships Investment Guide* (London: English Partnerships).

Evans, R. and Dawson, J. (1994), *Liveable Towns and Cities* (London: Civic Trust).

Falk, N. (1992), 'Turning the tide: British experience in regenerating urban docklands', in Hoyle, S. and Pinder, D.A. (eds.).

Fernie, K (2000), *Poverty, Disadvantage and Exclusion Profile 2000* (Dundee: University of Dundee School of Town and Regional Planning).

Fernie, K. and McCarthy, J. (2001), 'Partnership and Community Involvement: Institutional Morphing in Dundee', *Local Economy* 16:4, 299-311.

Fitzsimons, D. (1995), 'Planning and promotion: city reimaging in the 1980s and 1990s' in Neill, W. J. V., Fitzsimons, D. and Murtagh, B.

Gillett, E. (1983), *Investment in the Environment* (Aberdeen: Aberdeen University Press).

Gordon, G. (ed.) (1985) *Perspectives of the Scottish City* (Aberdeen: Aberdeen University Press).

Graham, S. (1995), 'The City Economy', in Healey P., Cameron, S., Davoudi, S., Graham, S. and Mandani-Pour, A (eds.).

Hall, S. (1995), 'The SRB: taking stock', *Planning Week*, 27.4.95, 16-17.

Hall, T. and Hubbard, P. (1998), 'The Entrepreneurial City and the "New Urban Politics", in Hall, T. and Hubbard, P. (eds.).

Hall, T. and Hubbard, P. (eds.) (1998), *The Entrepreneurial City: geographies of politics, regime and representation* (New York: John Wiley and Sons).

Harding, A. (1998), 'Public-Private Partnership in the UK', in Pierre, J. (ed.).

Hardy, S., Malbou, B. and Taverner, C. (eds.) (1996), *The Role of Arts and Sport in Local and Regional Economic Development* (London: Regional Studies Association).

Harvey, D. (1989), 'From managerialism to entrepreneurialism: the transformation in urban governance in late capitalism', *Geografiska Annaler* 71B, 3-17.

Harvey, D. (1997), 'Contested Cities: social process and spatial form', in Jewson, N. and MacGregor, S. (eds.).

Hastings. A. (1996), 'Unravelling the Process of Partnership in Urban Regeneration Policy', *Urban Studies* 33:2, 253-68.

Healey, P. (1996) 'Planning Through Debate: The Communicative Turn in Planning Theory', in Campbell, S. and Fainstein, S. (eds.).

Healey, P. (1997), *Collaborative Planning: shaping places in fragmented societies* (London: Macmillan).

Healey, P. (2006), *Urban Complexity and Spatial Strategies* (London: Taylor and Francis).

Healey, P., Cameron, S., Davoudi, S., Graham, S. and Mandanipour, A. (eds.) (1995), *Managing Cities: The New Urban Context* (London: John Wiley).

Healey, P., Cameron S., Davoudi, S., Graham, S. and Mandani-Pour, A. (1995), 'Introduction: The City – Crisis, Change and Intervention', in Healey, P., Cameron, S., Davoudi, S., Graham, S. and Mandanipour, A. (eds.).

Henderson, M. (1994), *Enterprise Zones*, Unpublished Research Project for the Degree in Town and Regional Planning (Dundee: Duncan of Jordanstone College, University of Dundee).

Hill, D.M. (2000), *Urban Policy and Politics in Britain* (London: Macmillan).

Hill, S. and Barlow, J. (1995), 'Single Regeneration Budget: Hope for 'those Inner Cities"? *Housing Review* 44 (2), March-April 1995, 32-35.

HMSO (1988), *Action for Cities* (London: HMSO).

Hoyle, S. and Pinder, D.A. (eds.) (1992), *European Port Cities in Transition* (London: Belhaven Press).

Hoyle, S., Pinder, D.A. and Husain, M.S. (eds.) (1988) *Revitalising the Waterfront: International Dimensions of Dockland Development* (London: Belhaven Press).

Imrie, R. and Thomas, H. (1993), *British Urban Policy and the Urban Development Corporation* (London: Paul Chapman Publishing).

Jacobs, B.D. (1992), *Fractured Cities: capitalism, community and empowerment in Britain and America* (London: Routledge).

Jacobs, B. (1992), *Fractured Cities: capitalism, community and empowerment in Britain and America* (London: Routledge).

Jessop, B. (1997), 'The Entrepreneurial City: re-imaging localities, redesigning economic governance, or restructuring capital?', in Jewson, N. and MacGregor, S. (eds.).

Jewson, N. and MacGregor, S. (1997), 'Transforming Cities: Social exclusion and the reinvention of partnership', in Jewson, N. and MacGregor, S. (eds.).

Jewson, N. and MacGregor, S. (eds.) (1997) *Transforming Cities: contested governance and new spatial divisions* (London: Routledge).

Jones S.J. (ed.) (1968), *Dundee and District,* (London: British Association for the Advancement of Science).

Judd, D. and Parkinson, M. (1990), (eds.). *Leadership and Urban Regeneration* (London: Sage Publications).

Judge, D. (1993), *The Parliamentary State* (London: Sage).

Kawashima, N. (1997), Local authorities and cultural policy: dynamics of recent developments, *Local Government Policy Making* 23, 31-39.

Keating, M. (1988), *The city that refused to die: Glasgow: the politics of urban regeneration* (Aberdeen: Aberdeen University Press).

Keating, M. and Boyle, R. (1986), *Remaking Urban Scotland: strategies for local economic development* (Edinburgh: Edinburgh University Press).

Kintrea, K., McGregor, A., McConnachie, M. and Urquhart, A. (1995), *Interim Evaluation of the Whitfield Partnership* (Edinburgh: Scottish Office Central Research Unit).

Kitchen, H. (1999) 'Turning community leadership into reality: a programme for new powers', in Local Government Information Unit (ed.).

Landry, C., Green, L., Matarasso, F. and Bianchini, F. (1996), *The Art of Regeneration: Urban Renewal through Cultural Activity* (Stroud: Comedia).

Law, C.M. (1988), 'Urban revitalisation, public policy and the redevelopment of redundant port zones: lessons from Baltimore and Manchester', in Hoyle, S., Pinder, D.A. and Husain, M.S. (eds)

Lawless, P. (1989), *Britain's Inner Cities* (London: Paul Chapman Publishing).

Le Gales, P. (2000) 'Private-sector interests and urban governance', in Bagnasco, A. and Le Gales, P. (eds.).

Lewis, O. (1968), 'The Culture of Poverty', in Moynihan, D. (ed.).

Lloyd, G. (2003) 'Scotland ties growth to its cities', *Regeneration and Renewal* 24 January, 14.

Lloyd, G. and Black, S. (1993), 'Property-led urban regeneration and local economic development', in Berry, J., McGreal, S. and Deddis, B. (eds.).

Lloyd, M.G. and McCarthy, J. (1999) 'Discovering Culture-led Regeneration in Dundee', *Local Economy* 14:3, 264-68.

Lloyd, M.G., McCarthy, J. and Peel, D. (2006), 'The re-construction of a small Scottish city', in Bell, D. and Jayne, M. (eds.).

Lloyd, M.G. and Newlands, D. (1988), 'The Growth Coalition and Uneven Economic Development', *Local Economy* 3:1, 31-39.

Local Government Information Unit (ed.) (1999), *Turning Community Leadership into Reality* (London: LGIU).

Lovering, J. (1997) 'Global Restructuring and Local Impact', in Pacione, M. (ed.).

Lowndes, V., Nanton, P., McCabe, A. and Skelcher, C. (1997), 'Networks, Partnerships and Urban Regeneration', *Local Economy* 11:4, 333-42.

Lyle, R. and Payne, G. (1950), *The Tay Valley Plan* (Dundee: Burns and Harris Ltd).

Mackintosh, M. (1992) 'Partnership: issues of policy and negotiation', *Local Economy* 7:2, 210-24.

Macmillan, D. (1999), 'A vision of brilliance on the banks of the Tay', *The Scotsman*, 22nd March, 11.

Mandanipour, A. (1998) 'Social Exclusion and Space', in Mandanipour, A., Cars, G. and Allen, J. (eds.).

Mandanipour, A., Cars, G., and Allen, J. (eds.) (1998), *Social Exclusion in European Cities: Processes, Experiences and Responses* (London: Jessica Kingsley).

Marcuse, P. and van Kempen, R. (2000a), 'Introduction', in Marcuse, P. and van Kempen, R. (eds.).

Marcuse, P. and van Kempen, R. (2000b) 'Conclusion: a changed spatial order', in Marcuse, P. and van Kempen, R. (eds.).

Marcuse, P. and van Kempen, R. (eds.) *Globalizing Cities: a new spatial order?* (London: Blackwell).

Massey, D.B. (1994), *Space, Place and Gender* (Cambridge: Polity Press).

Matarasso, F. (1997), *Use or Ornament? The Social Impact of Participation in the Arts* (Stroud: Comedia).

McArthur, A., Hastings, A. and McGregor, A. (1994), *An Evaluation of Community Involvement in the Whitfield Partnership* (Edinburgh: Scottish Office Central Research Unit).

McCarthy, J. (1995), 'The Dundee Waterfront: A Missed Opportunity for Planned Regeneration', *Land Use Policy* 12:4, 307-19.

McCarthy, J. (1998a), 'Waterfront Regeneration: Recent Practice in Dundee', *European Planning Studies* 6:6, 731-36.

McCarthy, J. (1998b), 'Dublin's Temple Bar - A Case Study of Culture-led Regeneration', *European Planning Studies* 6:3, 271-81.

McCarthy, J. (1998c), 'Reconstruction, Regeneration and Re-imaging. The case of Rotterdam', *Cities* 15:5, 337-44.

McCarthy, J. (2005), 'Promoting Image and Identity in Cultural Quarters: The Case of Dundee', *Local Economy* 20:3, 80-93.

McCarthy, J. (2006), 'The Application of Policy for Cultural Use Clustering', *European Planning Studies* 14:3, 397-408.

McCarthy, J. and Newlands, D. (1999) 'A Policy Agenda for the Scottish Parliament', in McCarthy, J. and Newlands, D. (eds.).

McCarthy, J. and Newlands, D. (eds.) (1999) *Governing Scotland: Problems and Prospects. The economic impact of the Scottish Parliament* (Aldershot: Ashgate).

McCarthy, J. and Pollock, S.H.A. (1997), 'Urban Regeneration in Glasgow and Dundee: a comparative evaluation', *Land Use Policy* 14:2, 137-49.

McCrone, G. (1991), 'Urban Renewal: The Scottish Experience', *Urban Studies* 28:6, 919-38.

Macdougall, D. (1993), 'Putting the heart Back in the City', in Duncan of Jordanstone College of Art.

McGregor, A. and McConnachie, M. (1998), 'Social Exclusion, Urban Regeneration and Economic Reintegration', *Urban Studies* 32:10, 1587-1600.

Molotch, H. (1976), 'The city as a growth machine. Towards a Political Economy of Space', *American Journal of Sociology* 82, 309-32.

Montgomery, J. (1995), 'The Story of Temple Bar: Creating Dublin's Cultural Quarter' *Planning Practice and Research* 10, 101-10.

Mooney, G. and Danson. M. (1997), 'Beyond "Culture City": Glasgow as a "dual city"', in Jewson, N. and MacGregor, S. (eds.).

Moynihan, D. (ed.) (1968), *Understanding Poverty* (New York: Basic Books).

Murray, C. (1984), *Losing Ground* (New York: Basic Books).

Myerscough, J. (1988), *The Economic Importance of the Arts in Britain,* PST, London.

Neill, W.J.V., Fitzsimons, D., and Murtagh, B. (1995), *Reimaging the Pariah City: Urban Development in Belfast and Detroit* (Avebury: Aldershot).

Nevin, B. and Beazley, M. (1995), 'Partnerships bridge divide but volunteers need boost', *Planning*, 28.4.95, 26-27.

Newlands, D., Danson, M. and McCarthy, J. (eds.) (2004), Divided Scotland. *The Nature, Causes and Consequences of Economic Disparities within Scotland* (Aldershot: Ashgate).

Oatley, N. (1995), 'Competitive Urban Policy and the Regeneration Game', *Town Planning Review* 66:1, 1-14.

Oatley, N. (1998), 'Cities, Economic Competition and Urban Policy', in Oatley, N. (ed.).

Oatley, N. (ed.) (1998), *Cities, Economic Competition and Urban Policy* (London: Paul Chapman Publishing).

Office of the Deputy Prime Minister (2003) *Sustainable communities: building for the future* (London: ODPM).

Office of the Deputy Prime Minister (2006), *An Exploratory assessment of the economic case for regeneration investment from a national perspective* (London: ODPM).

Osborne, D. and Gaebler, T. (1992), *Reinventing Government: How the Entrepreneurial Spirit is Transforming the Public Sector* (Cambridge, Mass: Addison-Wesley).

Pacione, M. (1985), 'Renewal, redevelopment and rehabilitation in Scottish cities, 1945-1981', in Gordon, G. (ed.).

Pacione, M. (1997), 'Urban Restructuring and the Reproduction of Inequality in Britain's Cities: An Overview', in Pacione, M. (ed.).

Pacione, M. (ed.) (1997), *Britain's Cities: Geographies of Division in Urban Britain,* (London: Routledge).

Paddison, R. (1993), 'City Marketing, Image Reconstruction and Urban Regeneration', *Urban Studies* 30:2, 339-49.

Parkinson, M. (1998), *Combating Social Exclusion: lessons from area-based programmes in Europe* (Bristol: The Policy Press).

Pierre, J. (ed.) (1998), *Partnerships in Urban Governance* (London: Macmillan).

Pocock, D.C.D. (1968), 'Dundee and its region', in Jones S.J. (ed.).

Roberts, P. (2004), 'Urban Regeneration in Scotland: Context, Contributions and Choices for the Future', in Newlands, D., Danson, M. and McCarthy, J. (eds.).

Robson, B. (1988), *Those Inner Cities* (Oxford: Clarendon Press).

Robson, B. (1994), 'Urban Policy - Patchy Past and a Fragile Future', *Municipal Review and AMA News*, Aug/Sept, 114-15.

Scottish Development Agency (1984), *Dundee Central Waterfront Development Brief* (Edinburgh: Scottish Development Agency).

Scottish Enterprise Tayside (2006), *Seabraes Yards. Dundee's Creative Media District* (Dundee: Scottish Enterprise Tayside).

Scottish Executive (2002a), *Better Communities in Scotland: Closing the Gap* (Edinburgh: Scottish Executive).

Scottish Executive (2002b) *Better Cities, New Challenges – A Review of Scotland's Cities* (Edinburgh: Scottish Executive).

Scottish Executive (2003a), *Local Government in Scotland Act Community Planning Statutory Guidance* (Edinburgh: Scottish Executive).

Scottish Executive (2003b), *A Partnership for a Better Scotland* (Edinburgh: Scottish Executive).

Scottish Executive (2003c), *Building Better Cities; Delivering Growth and Opportunities* (Edinburgh: Scottish Executive).

Scottish Executive (2004a) *A Framework for Economic Development* (Edinburgh: Scottish Executive).

Scottish Executive (2004b), *National Planning Framework for Scotland* (Edinburgh: Scottish Executive).

Scottish Executive (2004c), *Making Development Plans Deliver* (Edinburgh: Scottish Executive).

Scottish Executive (2006a), *People and Place: regeneration policy statement* (Edinburgh: Scottish Executive).

Scottish Executive (2006b), *Scottish Index of Multiple Deprivation* (Edinburgh: Scottish Executive).

Scottish Office (1988), *New Life for Urban Scotland* (Edinburgh: Scottish Office).

Scottish Office (1990), *Urban Scotland into the 1990s: new life two years on* (Edinburgh: Scottish Office).

Scottish Office (1993a), *Progress in Partnership: A Consultation Paper on the Future of Urban Regeneration Policy in Scotland* (Edinburgh: Scottish Office).

Scottish Office (1993b), *New Life for Urban Scotland: Second Monitoring Report to the [Whitfield] Partnership* (Edinburgh: Scottish Office).

Scottish Office (1995), *Programme for Partnership: Announcement of the Scottish Office Review of Urban Regeneration Policy* (Edinburgh: Scottish Office).

Scottish Urban Regeneration Forum (2006), *Reflect and Regenerate. Lessons from SURF's open forum programme* (Edinburgh: Communities Scotland).

Shiner, P. and Nevin, B. (1993), 'Helping the community take the driving seat in urban renewal', *Planning*, 12.11.93, 22-23.

Sinclair, D. (1997), 'Local government and a Scottish Parliament', *Scottish Affairs* 19, 14-21.

Smith, C. (1998), *Creative Britain* (London: Faber and Faber).

Smythe, H. (1994), *Marketing the City: the role of flagship developments in urban regeneration* (London: E and F N Spon).

Social Exclusion Unit (1998), *Bringing Britain Together – a national strategy for neighbourhood renewal* (London: Social Exclusion Unit).

Social Exclusion Unit (1991), *A New Commitment to Neighbourhood Renewal: National Strategy Action Plan* (London: Social Exclusion Unit).

Stone, C. (1989), *Regime Politics: Governing Atlanta 1946-1988* (Lawrence, Kansas: University of Kansas Press).

Subedi, Dr N.K. (1993), 'James Thomson', in Duncan of Jordanstone College of Art.

Taylor, C. A. (1990), *Local Planning and Waterfronts in Britain and the Netherlands*, Unpublished Dissertation for the Degree in Town and Regional Planning (Dundee: Duncan of Jordanstone College, University of Dundee).

Tiesdell, S., Adams, D., Hastings, A. and Turok, I. (2006), 'A step-change in Scottish Regeneration', *Town and Country Planning*, April, 104-07.

Thorsby, D. (1994), 'The production and consumption of the arts: a view of cultural economics', *Journal of Economic Literature* 32, 1-29.

Tunbridge, J. (1988), 'Policy convergence on the waterfront? A comparative assessment of North American revitalisation strategies', in Hoyle, S., Pinder, D.A. and Husain, M.S., (eds.).

Tunstall, R. and Coulter, A. (2006), *Twenty-five years on twenty estates* (Bristol: Policy Press).

Turok, I. (1987), 'Continuity, change and contradiction in urban policy', in Donnison, D. and Middleton, A. (eds.).

Turok, I. (1992), 'Property-led urban regeneration: panacea or placebo?', *Environment and Planning A*, 24, 361-79.

University of Dundee Press Office (1999), 'DCA opens at last', *Contact* 16, March, 14-15.

Urban Task Force (1999), *Towards an Urban Renaissance: Final Report of the Urban Task Force chaired by Lord Rogers of Riverside* (London: E and F N Spon).

Urban Task Force (2005), *Towards a Strong Urban Renaissance* (London: Urban Task Force).

Valler, D. (1995), 'Local economic strategy and local coalition building', *Local Economy* 10:1, 33-47.

Vigar, G., Healey, P., Hull, A. and Davoudi, S. (2000), *Planning, governance and spatial strategy in Britain* (London: Macmillan).

Wannop, U. (1995), *The Regional Imperative. Regional Planning and Governance in Britain, Europe and the United States* (London: Jessica Kingsley Publishers).

Wannop, U. and Leclerc, R. (1987), 'Urban renewal and the origins of GEAR', in Donnison, D. and Middleton, M. (eds.).

Ward, S.V. (1994), *Planning and Urban Change* (London: Paul Chapman Publishing Ltd).

Whatley, C.A. (1991), 'The making of 'Juteopolis' - and how it was', in Whatley, C.A. (ed.).

Whatley, C.A. (ed.). (1991), *The Remaking of Juteopolis*, Proceedings of the Abertay Historical Society's Octocentenary Conference (Dundee: Abertay Historical Society).

Whitfield Partnership (1988), *Whitfield Partnership: First Report on Strategy* (Dundee: Whitfield Partnership).

Whitfield Partnership (1989), *Baseline Report: Spring 1989* (Dundee: Whitfield Partnership).

Williams, C. (1996), 'An evaluation of approaches towards cultural industries in local economic development: a comparison of Sheffield, Glasgow and Leeds' in Hardy, S., Malbou, B. and Taverner, C. (eds.).

Williams, C. (1997), *Consumer Services and Economic Development* (London: Routledge).

Wood, A. (1996), 'Analysing the politics of local economic development: making sense of cross-national convergence', *Urban Studies* 33, 1281-95.

Index